A PRACTICAL APPROACH TO RELIABILITY

By the same author

A Practical Approach to Quality Control

A PRACTICAL APPROACH TO RELIABILITY

Rowland Caplen
B.Sc., C.Eng., M.I.E.E., M.Inst.P., F.I.E.I.

BUSINESS BOOKS LIMITED
London

ISBN 0 220 66809 4

This book has been set in 11 on 13pt Baskerville: printed in England by Clarke, Doble & Brendon Ltd, Plymouth for the publishers Business Books Limited (registered office: 180 Fleet Street, London, E.C.4) publishing offices: Mercury House, Waterloo Road, London, S.E.1.

Made and printed in Great Britain

CONTENTS

LIST OF FIGURES

INTRODUCTION

This book has been written to meet the needs of the practical man, who finds himself limited to either complicated theoretical works, or to elementary texts. The author has two main objectives:

1 To provide a practical introduction to reliability, which will enable people in both manufacturing and service organisations, to appreciate its possibilities in their field.
2 To satisfy the needs of those studying reliability for the examinations of the City and Guilds of London Institute and the Institution of Engineering Inspection.

For completeness essential theory has been included, but every attempt has been made to avoid unnecessary mathematics, and to simplify what remains. While some knowledge of algebra and calculus is advantageous, no prior knowledge of statistics has been assumed, and most of the book is quite readable to the non-mathematician.

Terms are in accordance with British Standard Specification 4200 as far as they are available.

This book is a companion volume to *A Practical Approach to Quality Control*, by the same author.

Spring 1972 ROWLAND CAPLEN

PART ONE **RELIABILITY PRINCIPLES AND COSTS**

CHAPTER 1 AN INTRODUCTION TO RELIABILITY

1.1. reliability in the past

Suppose we begin by considering something which has proved its reliability over the years. In 1826 Thomas Telford completed the suspension bridge across the Menai Strait in North Wales. Since then it has been used extensively, and it still carries all road traffic to and from the Isle of Anglesey. Now Telford could not have known anything about reliability in the modern sense, yet his design proved to be very successful. Probably two factors assisted him:

1. The bridge is a relatively simple structure, and although it is large there is not a lot to go wrong.
2. It is probably much stronger than is really necessary. The extra weight and size were unlikely to matter much, and the additional cost was presumably acceptable.

Many factors were no doubt left to chance. Thus if Telford had got his safety factors all wrong, the bridge might have fallen down within a short time and by now be almost forgotten. Hence old structures which still survive do so because they were reliably constructed in the first place. Those which were not have long since gone, and this sometimes leads people to say, not necessarily correctly, 'Ah, they don't make things as well as that today'.

1.2. definition of reliability

Next consider the reliability of a motor car. If it always work perfectly, we say that it is 100 per cent reliable, whereas if it sometimes breaks down, its reliability is less than 100 per cent. Suppose however that on investigating a breakdown, we discover that it is a very old car. We shall conclude that it is worn out. Since nothing lasts for ever, reliability can only be meaningful if we relate it to the time we expect our car to last. A new car might be very reliable if we only expect it to last for 5 years, rather less reliable over a period of 10 years, and completely unreliable if we are expecting a useful life of say 40 years. Thus a figure for reliability is meaningless unless we also state the planned working life.

Similarly we shall need to know the intended conditions of use, and the routine maintenance which is required. If a car engine seizes up because there is no water in the radiator, we shall regard this as a failure of maintenance rather than a failure of reliability. If a car fails when its ambitious driver is trying to set up a record for driving across the Sahara desert, we shall say that this is a misuse failure.

We can now collect all these points together and ask whether, if a particular car is used as intended and maintained according to the manufacturer's instructions, it will be reliable for say 10 years. Unfortunately there is only one certain way to find out and that is to run it for 10 years and see. If at the end of that time it is still working satisfactorily, then it was a reliable car. The snag is that by then it is worn out, and anyway we have had to wait 10 years for the information.

One reason we cannot predict at the start whether any particular car will be reliable is that no two are ever absolutely identical, even though they are the same make and model. There are always small manufacturing differences, and a few may contain defects. The car designer will try to make his car insensitive to the likely variations, and the factory quality engineer will try to reduce them, and eliminate any defects. If both were completely successful, and if every car was used identically, then every car would have the same reliability. In practice this is impossible.

Suppose that out of every 100 cars of a particular type, 99 prove to be trouble free if used and maintained correctly, and one fails to work as intended. Then we can say that the reliability of each car is 99 per cent, meaning that the chances are 99 in 100 that it will prove reliable. For mathematical reasons, we usually express reliability in decimals of 1·00 as follows:

Reliability = 1·00 means certain to work as intended,
Reliability = 0·99 means 99 per cent likely to work as intended,
Reliability = 0·50 means 50 per cent likely to work as intended,
Reliability = 0·00 means absolutely certain not to work as intended.

We can now give reliability a formal definition as follows. *Reliability* is the probability that an item will perform as required, under stated conditions, for a stated period of time. Thus if we have a large number of items on test, we can write :

$$\text{Reliability at time } t = R(t) = \frac{\text{Number surviving at instant } t}{\text{Number at start when } t=0} \qquad (1.2a)$$

1.3. definition of failure

When an item no longer works as intended we say it has failed, and hence as a definition :
Failure is the termination of the ability of an item to perform its required function.

This however assumes that we know exactly what its required function is. Thus in the last section, we assumed that a car either works perfectly or breaks down completely. Suppose however that due to some mechanical defect, it will not go faster than 60 kilometres per hour. Do we consider this a failure or not? It depends entirely on what we decide is 'performing its required function', and this in turn may depend on the application.

Suppose we have a circuit which requires a 100 Ω resistor, and the resistor allocated for the job happens to have a resistance of 102 Ω. If the application will accept ± 5 Ω this will be all right, but if it will only accept ± 1 Ω, that same resistor will count as a failure.

1.4. classification of failures

It is convenient to classify failures as follows.

1 AS TO CAUSE

A misuse failure is, as a definition, a 'failure attributable to the application of stresses beyond the stated capability of the item'. Thus it has been ill treated.

An inherent weakness failure is a 'failure attributable to weakness inherent in the item itself, when subjected to stresses within the stated capabilities of the item'. Thus the item has not been ill treated, and its failure is probably due to a design or manufacturing fault.

2 AS TO SUDDENNESS

A sudden failure is one which could not be anticipated by prior examination.

A gradual failure is one which could be anticipated by prior examination. Thus because it takes place gradually it is possible to predict that it will occur.

3 AS TO DEGREE

A partial failure is one resulting 'from deviations in characteristic(s) beyond specified limits, but not such as to cause complete lack of the required function'. Thus the item does not work as well as it should, but it has not completely failed.

A complete failure is one resulting from 'deviations in characteristic(s) beyond specified limits, such as to cause complete lack of the required function. The limits referred to in this category are special limits for this purpose'.

4 BY COMBINATION OF THE ABOVE TERMS

A catastrophic failure is one which is both sudden and complete.

A degradation failure is one which is both gradual and partial.

1.5. the reduction of reliability with time

In order to observe reliability in practice, suppose that from a large batch of some particular part, we take a random sample of 1000, check to confirm they work satisfactorily and put them on life test. By this we mean that we shall work them as nearly as possible as they would be worked in practice, and we shall note when each part fails. If we decide not to replace failures with good parts, the number remaining on test will gradually reduce.

Suppose that exactly 100 hours after the start of the test we notice that 221 parts have failed, leaving 779 still working. Then from section 1.2 we have:

$$\text{Observed reliability over 100 hours} = \frac{\text{Number surviving at time } t = 100}{\text{Number at the start of the test}}$$

$$= \frac{779}{1000}$$

$$= 0{\cdot}779$$

Notice that reliability is a cumulative quantity. In order to work out the number surviving after 100 hours, we have in effect to add up all the parts which have failed so far and take the total from the original 1000. Thus it follows that for any other period of time the reliability would be different. After the test had run

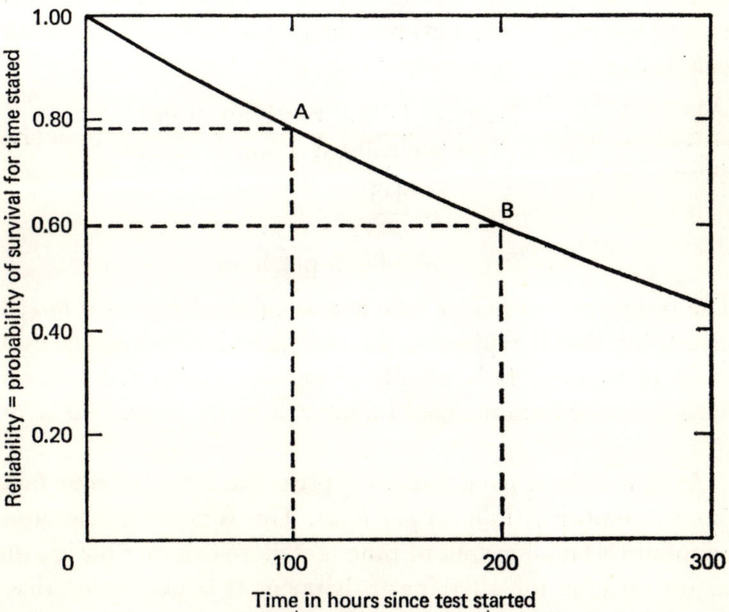

FIG 1.5(a). The relation between reliability and time in operation

200 hours the number of failures might have risen to 400, and then we should have:

$$\text{Observed reliability over 200 hours} = \frac{600}{1000} = 0{\cdot}600$$

It is nearly always convenient to count the start of the test as time $t = 0$, so we are cumulating all the failures from then until the moments when time t reaches 100 and 200 hours, respectively. Indeed we could plot a graph of reliability against time, as shown in Fig. 1.5(a). Notice how the reliability progressively falls, telling us what we already know, that the longer we expect anything to last, the more likely it is to fail during that time.

1.6. failure rate

Suppose that at the instant when $t = 200$ hours, we are able to find out that parts are failing at exactly 1·5 per hour. Since 600 are still working, we can express the rate at which they are failing as follows:

$$\text{Observed instantaneous failure rate} = \frac{\text{Number failing per hour at instant } t}{\text{Number still surviving at instant } t} \qquad (1.6\text{a})$$

$$= \frac{1{\cdot}5}{600}$$

$$= 0{\cdot}0025 \text{ per hour}$$

The *instantaneous failure rate* is also called the *hazard rate*, the idea being that it represents the rate at which we risk the occurrence of failures. It is usually given the symbol z, so that the hazard rate or instantaneous failure rate at the instant t is written $z(t)$.

At first sight it is curious to express the *instantaneous* failure rate in number of failures *per hour*. This is necessary because in an infinitely small instant of time no failure can ever occur; there is just no time for anything to happen. It is like saying that at exactly 10.00 hours this morning (an instant in time) we were travelling at precisely 80 kilometres per hour. There is no implication that we actually went as far as 80 kilometres, or that we motored for a whole hour. It merely expresses our rate of change of distance with time, at that instant.

The instantaneous failure rate is a mathematical concept, for clearly in our practical example, we would not know that at that instant, failures were occurring at exactly 1·5 per hour. In practice we must observe failures over a finite period of time. If as is often the case the failure rate happens to be constant, this makes no difference to the result. If it varies, then the figure we obtain will be the *mean failure rate* over that period.

Where the failure rate is constant, or is averaged, it is usual to denote it by the Greek letter lambda which is written λ. Therefore:

$$z(t) \text{ becomes } \lambda$$

Indeed reliability engineers are not completely agreed about failure rate symbols. Since many practical failure rates are either constant or averaged, the symbol λ is frequently used, and some engineers write λ or $\lambda(t)$ for what we have called $z(t)$.

We can now define *failure rate* as the probability of failure in unit time of an item which is still working satisfactorily. By 'still working satisfactorily' we mean that we shall not count failures which have occurred before the instant of time in which we are interested. Thus in our calculation the denominator is 600 and not 1000, because we did not count the 400 which failed prior to the moment when the time t was 200 hours.

1.7. the failure probability density function

Obviously we could have based the above discussion on the 1000 parts which originally started the test. Having noted how many parts failed during each hour of the test, we could have said that x_1 per cent of the 1000 failed during the first hour, x_2 per cent during the second hour, and so on up to say, x_n per cent during the hour when the 1000th part failed. Since all the parts have now failed, if we add x_1+x_2+, etc. up to x_n, the total must be 100 per cent. As with values of reliability, we usually express the proportion failing as a decimal of 1·00, rather than as a percentage. Hence if we add all the values together, the total will be 1·00 instead of 100 per cent.

There is one further refinement. We considered the fraction which failed during each successive hour, but the failure rate may vary over a period as long as that. It would be more accurate to take periods of say 30 minutes, but even this would not be as good as periods of 15 minutes, and so on. Therefore for absolute accuracy, the statistician will prefer infinitely small periods of time. This can be done if we express the *probability of failure* as a mathematical *function*, such as an equation. The statistician will say that such an equation shows the *density* with which failures are likely to occur at any instant, and it is therefore called the *failure probability density function.*

Thus, as a definition, the failure probability density function gives, for any instant of time t, the probability of failure in unit time of an item which was working satisfactorily at time $t = 0$. In our example, parts were failing at exactly 1·5 per hour at the instant when $t = 200$ hours, so we can write:

$$\text{Observed value of failure probability density function for instant } t = \frac{\text{Number failing per hour at instant } t}{\text{Number at start } (t=0)} \quad (1.7a)$$

$$= \frac{1{\cdot}5}{1000}$$

$$= 0{\cdot}0015 \text{ per hour}$$

Notice that as with the instantaneous failure rate, the failure probability density function is expressed in failures per unit time, usually 1 hour, even though it is considering an instant in time.

The symbol $f(t)$ is used to denote the failure probability density function. We have already said that if we add up the fraction which fail for every instant of time, the total must be 1·00. For a continuous mathematical function we can write this as

$$\int_{t=0}^{t=\infty} f(t) = 1{\cdot}00 \quad (1.7b)$$

Mathematically we must integrate from $t = 0$, when the test starts, to $t =$ infinity. In practice the total must be 1·00 when every part on test has failed.

1.8. the relation between *R(t)*, *z(t)* and *f(t)*

$R(t)$, $z(t)$ and $f(t)$ are closely related. Thus from equations (1.3a) and (1.6a) we have:

$$R(t) \times z(t) =$$

$$\left[\frac{\text{Number surviving at instant } t}{\text{Number at start } (t=0)}\right] \times \left[\frac{\begin{array}{c}\text{Number failing per hour}\\ \text{at instant } t\end{array}}{\begin{array}{c}\text{Number surviving at}\\ \text{instant } t\end{array}}\right]$$

$$= \frac{\text{Number failing per hour at instant } t}{\text{Number at start } (t=0)}$$

$$= f(t) \text{ from equation (1.7a)}$$

Therefore: $$f(t) = R(t) \times z(t) \tag{1.8a}$$

Alternatively, we can derive this result as follows.
From the definition in section 1.7 we have: $f(t) =$ probability of failure at some instant of time t, of an item which was satisfactory at the start, when $t = 0$. Now in order to fail at time t, and not before, an item starting at time $t = 0$ must:

1 Survive to time t, and then
2 Fail at time t.

But the probability it will survive to time t is its reliability expressed by $R(t)$, and the probability that having survived to time t it will then fail at that instant is given by the instantaneous failure rate $z(t)$.

$$\text{Therefore } f(t) = \left[\begin{array}{c}\text{Probability of survival}\\ \text{to time } t\end{array}\right] \times \left[\begin{array}{c}\text{Probability of failure}\\ \text{at instant } t\end{array}\right]$$

$$\text{Therefore } f(t) = R(t) \times z(t) \tag{1.8a}$$

Thus in our example we had, for the instant when $t = 200$ hours,

$$R(t) = 0{\cdot}60$$
$$z(t) = 0{\cdot}0025 \text{ failures per hour}$$

Therefore:

$$f(t) = R(t) \times z(t)$$
$$= 0{\cdot}60 \times 0{\cdot}0025$$
$$= 0{\cdot}0015 \text{ failures per hour}$$

This agrees with our calculation in section 1.7.

CHAPTER 2 FAILURE PATTERNS

2.1. the bath tub curve

In section 1.6 we noted that the failure rate is not necessarily constant, and we must now consider how it may vary. Suppose we put a large number of a particular part on life test, and we were able to run the test until every part has failed. We note when each failure occurs, although we do not replace the part, and we then plot a graph of observed failure rate against time since the test started. Fig. 2.1(a) shows a typical result we might get. Because of its shape, this is called the bath-tub curve, and it divides naturally into three periods as follows.

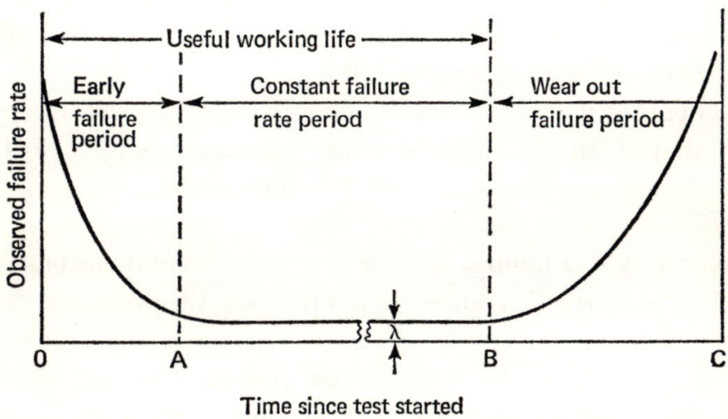

FIG 2.1(a). Typical bath-tub curve of failure rate against time

2.2. early failure period

At the start of our test the failure rate may be relatively high, but this usually falls progressively, until at A we enter a period where the failure rate is approximately constant and at its lowest level. We have all had experience of buying something new, and then in the first few hours or weeks we have trouble with it. It is the period often covered by guarantee, during which the manufacturer agrees to make good anything which goes wrong.

The most common causes of early failures are:

1 MANUFACTURING FAULTS These are faults which were not detected before despatch to the customer. In each case, two faults are implied:

a The product was wrongly made.
b Its fault was not detected before it left the factory.

Manufacturing faults often account for the majority of the early failures.

2 DESIGN FAULTS When a designer first completes a new design, it must be thoroughly tested as a prototype before full scale manufacture begins. If this is not adequately done, then the customer is in effect asked to do the manufacturer's prototype tests as part of normal use. Any shortcomings in the design may then reveal themselves as early failures.

Sometimes a design may function so near to the acceptable limit that if all the manufacturing tolerances happen to be in unfavourable directions, an overall failure results.

3 MISUSE A few failures may be due to accidental customer misuse, before he is fully competent in operating the product.

Whatever the cause, apart from customer misuse, the fault must be corrected as quickly as possible, with a minimum of inconvenience and no cost to the customer. The good name of the manufacturing company is at stake, and the money loss resulting from a scatter of dissatisfied customers may be very considerable. Statistics here are no protection. It is no good telling a housewife who has bought a brand new but faulty washing machine that, 'only 1 in 10,000 of our washing machines is faulty'. As far as she is concerned, she has bought one machine, and it is faulty and that is 100 per cent wrong.

2.3. constant failure period

Once the early failures have been removed, the parts usually settle down to what may be a relatively long period, when the failure rate is approximately constant. In Fig. 2.1(a) this continues from A to B, after which the failure rate begins to rise again, often quite steeply as the parts begin to wear out. Thus the two periods from O to B make up the useful working life of the parts, and to avoid failures we must replace just before time B when wear out failures start.

Although the failure rate in the constant failure period is usually low, it can be very troublesome if high reliability is required. We can avoid early failures by good design and manufacture, and by running the parts on load for a time at least equal to OA of Fig. 2.1(a). We can avoid wear out failures by replacement before time B, but we are still left with the constant failure rate period, right through the normal working life.

These failures are unlikely to be due to any single cause, although if this should be so we have an immediate clue to what is wrong. If for example one particular area of a component always overheats, then we can investigate why this should be. Failures which are due to a single cause, however, are unlikely to occur at a constant rate over a long period. It is much more likely that all the failures from a particular cause will tend to occur somewhere about the same time, or with a well defined pattern.

Thus during the constant failure period, it is usual for failures from a wide variety of causes to occur at random, with no obvious pattern, except that the failure rate is roughly constant. Statistically they are said to be random or chance failures, but these terms are unfortunate because they imply that nothing can be done about them. They are only chance failures because we do not know the reason for them, and because without any well defined pattern the causes of failure will be difficult to determine and correct.

Where failures do form a well defined pattern, the reliability engineer calls them *systematic failures*, and such patterns usually provide valuable information about the cause of failure.

2.4. the wear out failure period

Everything wears out sooner or later, and so after B in Fig. 2.1(a), the failure rate rises again. The wear out situation is better seen if we plot, as in Fig. 2.4(a), the corresponding graph of the failure probability density function and compare it with the failure rate.

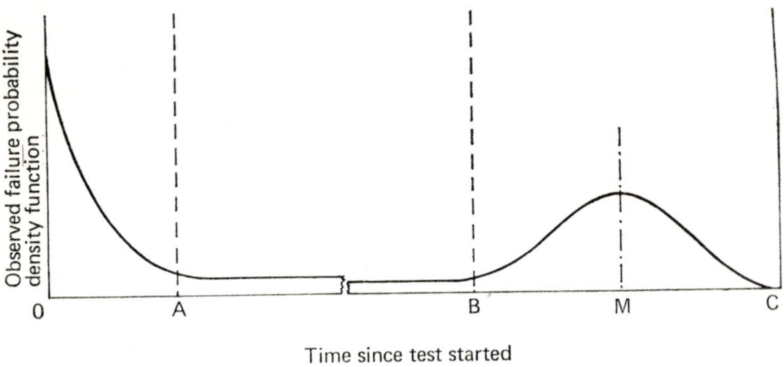

FIG 2.4(a). Observed values of the failure probability density function against time

Both curves start off with the same value. This must be so, because we have assumed that all the parts work satisfactorily at the moment the test starts.

Thus reliability for $t = 0$ is

$$R(t_0) = 1{\cdot}0$$

and inserting this in equation (1.8a) we get

$$f(t_0) = z(t_0)$$

During the early failure period, the failure probability density function drops a little more quickly than the failure rate, and during the constant failure rate period it is not quite constant. It reduces gradually as the number of parts on test reduces.

It is in the wear out period however that the difference is

really marked, because $f(t)$ rises to a peak at M and then falls again to a very low value. The reason is that although the failure rate (i.e. the probability of failure of parts still working) is increasing rapidly, the actual number of parts still working is reducing rapidly. Therefore, beyond C very few parts fail, because most of them have already failed.

Sometimes a reasonably well shaped symmetrical curve results as shown in BMC of Fig. 2.4(a), in which case M is the mean time to wear out failure, and parts are just as likely to fail from wear out a little before this time, as to last a little longer beyond M.

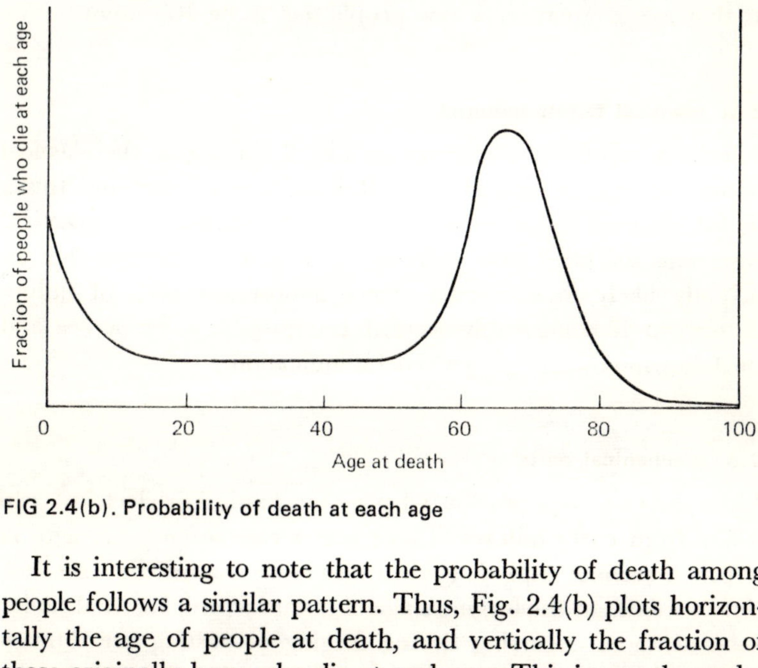

FIG 2.4(b). Probability of death at each age

It is interesting to note that the probability of death among people follows a similar pattern. Thus, Fig. 2.4(b) plots horizontally the age of people at death, and vertically the fraction of those originally born who die at each age. This is exactly analogous to the failure probability density function, which plots the fraction of parts originally put on test or into service against the age at which failure occurs.

Some babies are unlucky enough to be born with physical weaknesses, and so die in the first few years of life. This is the infant mortality period, and it corresponds to the early failure period. Indeed, some reliability engineers refer to the early failure period as the infant mortality period.

After the age of about 10 years, the risk of death from natural causes drops to its lowest level, and the majority of those who survive the first few years of life will live for some while longer. In the fifties and sixties, however, the curve rises as people begin to die of old age, which is of course the equivalent of wear out in the reliability sense. Beyond about 70 years it falls again, because although the risk of death is increasing, only a small fraction of those originally born are still alive. Thus, few people die at the age of 100 years because, although the risk of death at that age is great, very few people live to be 100 anyway.

2.5. practical failure patterns

The bath tub curve illustrated in Fig. 2.1(a) shows the classical pattern of failure against time. Reliability engineers find this a useful starting point from which to investigate any practical case, and we must now consider what sort of patterns we are actually likely to encounter. Since almost any type of failure pattern could conceivably occur, it is impossible to be precise and the following merely gives a general indication.

2.6. mechanical parts

Most parts and equipments, whether mechanical or electrical, can suffer from early failures. There was a case some years ago, in which a turbine exploded when one of its blades disintegrated, due to a fault at the forging stage of manufacture. Internal faults, such as those occurring during forging or casting are difficult to eliminate completely unless extensive testing is carried out.

Mechanical parts can be loosely considered to start the gradual process of wearing out more or less from the moment they are put into service. Certainly wear out starts as soon as wear in is complete. Thus, wear continues, not necessarily at a constant rate, throughout the working life, and sooner or later it will be sufficient to cause a wear out failure. Therefore, many mechanical parts exhibit some sort of wear out failure period, if they are worked for long enough. In some cases this may start so soon that there is little, if any, constant failure period. In others it may be too late to have any practical significance. For example the frame of a machine tool will usually outlast the moving parts.

If the bell shape of the wear out failure period extends over an excessively long time, this suggests that there have been considerable variations in the quality of manufacture, or if the curve is based on field experience it may be due to large variations in use and treatment by customers.

2.7. electrical and electronic parts

Some products such as switches, relays, electric motors and generators which we probably think of as electrical, have mechanical moving parts. Often the electrical side will outlast the mechanical, and so the failure rate will be more akin to the mechanical case.

Other parts, while essentially electrical, nevertheless exhibit a wear out pattern. Dry cell batteries for example suffer from progressive chemical changes, and so eventually fail. Incandescent lamps and vacuum tubes also often show a pronounced wear out pattern.

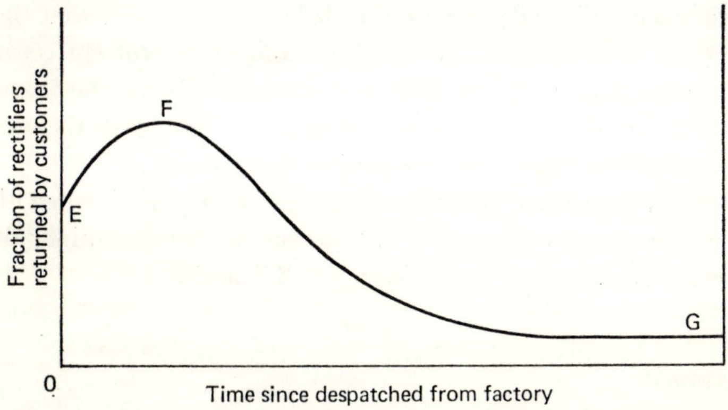

FIG 2.7(a). Fraction of rectifiers returned by customers, against time since they were despatched

At the other extreme, some electronic parts suffer almost no progressive change, and so may have a very long life. After the removal of any early failures, they settle down to a constant failure rate, which is so long that wear out failure never seems to occur. Many semi-conductor rectifiers and transistors are in this class. Some of the earliest copper oxide rectifiers have now been in use for over 40 years, and there seems to be no indication that they will ever wear out. From time to time they are discarded because their output is too small for modern requirements, and they will of course fail if excessively overloaded.

If we make an attempt to plot a bath tub curve using information about items returned by customers as unsatisfactory, we shall probably involve inaccuracies due to our lack of knowledge about timing. Thus, Fig. 2.7(a) shows a typical graph which a manufacturer of semiconductor rectifiers might obtain if he plots the fraction of his despatches which are eventually returned by customers as faulty, against the time when they arrive back.

At the start EF, the curve rises a little before it falls in the way we should expect. This is usually because the manufacturer only knows the date of despatch, and it may be some time after this before the customer puts the product into service.

There is no wear out section corresponding to BC of Fig. 2.4(a). Since the life of semiconductor rectifiers is very long, when they do eventually fail the customer is likely to consider that they have given a satisfactory working life, and so he will not return them to the manufacturer with a complaint. Hence the manufacturer never receives the necessary information to plot the wear out section of the curve, even supposing that it exists.

So many electronic parts have a very long period when the failure rate is approximately constant that we must consider this important case further in sections 3.2, 3.3 and 3.4.

2.8. systems

Various terms are used to describe equipments. They are not precisely defined, and their meaning varies according to the size of the equipment. Thus the following is merely a general guide.

1 *System* is used to denote any complete installation or equipment. Thus a system might be gigantic, such as a large radar system, but it might equally be quite small.
2 A system can then be divided into *blocks* each of which:
a Performs an integrated function.
b Can if desired be divided into units.
3 In a large system, even a *unit* may be quite a sizeable subsystem, but at the other extreme it may be merely a circuit containing a number of components.
4 The unit can be broken down into *elements*, each of which may be a circuit or just a component.
5 *Components* and *parts* may be the same thing, but sometimes a component contains a number of parts.
6 The word *item* is quite general, and may refer to a system, block, unit, element, component or part.

As a first approximation, the failure pattern of a system, or indeed any assembly of parts, can be regarded as the sum of the failure patterns of the individual parts, but there may be additional failures as follows:

1 Failures may occur at what are termed the *interfaces* between two parts. For example there may be failures in soldered joints or connectors, as well as in the parts themselves.
2 One part may affect the performance of another. Thus if one part fails, it may overload other perfectly good parts and cause them to fail as well.

The above suggests that system failure rates will be difficult to deal with, but fortunately this is not necessarily so, as we can see from the following analogy. If we were to listen to one person throughout the day, his speaking pattern would be irregular. Sometimes he would talk loudly, sometimes softly and sometimes not at all. However, if we go into our factory canteen, where lots of people are talking, our general impression is of a roughly constant noise level, because the loud and soft mingle together.

Return now to our failure rate problem, and suppose that we operate a group of similar systems, each containing a large number of parts. Whenever a system fails we repair it, probably by replacing the faulty part with a new one. When repair is no longer economic, we buy a new system. Thus, each system will be a different age, and each of the large number of parts it contains will have its own failure pattern. The addition of so many small failure patterns is likely to produce a roughly constant failure rate for the whole system, in exactly the same way as the noise level in our canteen was roughly constant.

If, for example, we keep a fleet of 200 delivery vans, which we repair and replace as necessary, we should expect our failure rate and so our maintenance costs to be roughly constant. Of course, if we bought 200 new vans and started using them all together, then we should expect an initial crop of early failures, before settling down to a roughly constant failure rate.

2.9. mean time between failures

Where the failure rate is approximately constant, it is often convenient to use the *mean time between failures* (MTBF). Thus suppose a particular part has a constant failure rate of 0·002 per hour. If we ran a test on one part, replacing it every time it failed, we should expect it to fail twice per 1000 hours on average, which is once per 500 hours. Therefore: Mean time between failures = 500 hours.

The word 'expect' is used in the statistical sense. Although we expect failures to occur once per 500 hours on average, we know that the time between any two actual failures might vary considerably from this. From the above it follows that the mean time between failures is the reciprocal of the failure rate. Thus let

Mean time between failures $= \theta$
Constant failure rate $= \lambda$

$$\text{Therefore } \theta = \frac{1}{\lambda} \qquad (2.9a)$$

In the above example

$$\text{MTBF} = \theta = \frac{1}{0{\cdot}002}$$
$$= 500 \text{ hours}$$

As we have seen, there are many cases, especially with electronic parts and with complete systems, where the failure rate is approximately constant over the useful working life. In such cases it is common to quote the MTBF rather than the reliability in contracts and specifications. Reliability has the disadvantage that it always relates to a particular planned working life, as explained in section 1.2, and this may or may not be the useful working life span the customer requires. Mean time between failures is constant, regardless of the life span, provided that it is not so long that it runs into the wear out failure period. If we know the mean time between failures, we can easily calculate the reliability for any desired life span, and we shall discuss how to do this in section 3.2.

2.10. mean time to failure

The term mean time to failure (MTTF) is analogous in every way with mean time between failures, and is used where a failure cannot be repaired. It would for example be used for the parts of an artificial satellite, where if a failure occurs after launch little can usually be done about it.

CHAPTER 3 FITTING CURVES TO FAILURE PATTERNS

3.1. relating theory with practice

Where the failure pattern closely fits a known algebraic equation, then this equation may be useful in order to predict what will happen in other similar situations. Thus if we have a known constant failure rate, we can calculate the reliability over any period we choose. Considerable caution is however necessary with a prediction of this sort, because we are not usually dealing with an exact scientific relationship, and unless the equation we use really fits our particular problem, the prediction may be valueless.

3.2. calculation of reliability when the failure rate is constant

Consider first the important case where the failure rate is approximately constant. Now, as we said in section 1.5, reliability is cumulative. When we say that the reliability of 100 parts over the period from time 0 to time t_1 was observed to be 0·70, we mean that if we *add up* all those which failed between the start of the test and time t_1, there would be a total of 30 failures out of 100.

Suppose we continued the test to a new time t_2, and during this extension a further 3 parts failed, so that the reliability is now 0·67, therefore:

Observed change in reliability from time t_1 to t_2 = (New reliability) − (Previous reliability)

$$= 0{\cdot}67 - 0{\cdot}70$$
$$= -0{\cdot}03$$

The negative sign means that the reliability is reduced. But from section 1.7, the failure probability density function tells us the fraction of the original number of parts which fail at each instant of time. Therefore over the period t_1 to t_2 for which we extended the test, if we add together the fraction failing for all these 'instants' the total must be 0·03, because 3 per cent of the original parts failed in this period. Thus at any one instant:

The value of the failure probability density function = − (The rate of change of reliability)

Writing this in symbols, and relating it to a very short time interval dt, we get:

$$f(t) = \frac{-\mathrm{d}R(t)}{\mathrm{d}t} \tag{3.2a}$$

but from equation (1.8a) we already have:

$$f(t) = R(t) \times z(t)$$

Therefore

$$\frac{-\mathrm{d}R(t)}{\mathrm{d}t} = R(t) \times z(t)$$

We rearrange the terms to give:

$$\frac{\mathrm{d}R(t)}{R(t)} = -z(t)\mathrm{d}t$$

Next we integrate from time $t = 0$ when the test started to time t, say, when it finished. Therefore:

$$\int \frac{\mathrm{d}R(t)}{R(t)} = \int_0^t -z(t)\mathrm{d}t$$

So far our argument has been quite general, and we have not assumed that $z(t)$ is constant. Now however we cannot integrate unless $z(t)$ is either constant or we know its pattern of variation. Assuming therefore that $z(t)$ is constant, it becomes λ as explained in section 1.6, and we have:

$$\log R(t) = -\lambda t$$

and taking antilogs,

$$R(t) = \exp(-\lambda t) \tag{3.2b}$$

Exp$(-\lambda t)$ can also be written $e^{-\lambda t}$ since exp stands for the exponential constant = 2·71828. Since the log of $\exp(-\lambda t)$ is $-\lambda t$, it follows as above that the antilog of $(-\lambda t)$ must be $\exp(-\lambda t)$.

To calculate the failure probability density function $f(t)$, we use equation (1.8a).

Therefore:
$$f(t) = z(t) \times R(t)$$
$$= \lambda \times R(t)$$

Therefore using (3.2b)
$$f(t) = \lambda \exp(-\lambda t) \tag{3.2c}$$

EXAMPLE

Suppose that a part has a constant failure rate of 0·001 per hour, and we wish to calculate its reliability over 500 operating hours. From equation (3.2b)

$$\begin{aligned} R(t) &= \exp(-\lambda t) \\ &= \exp(-0{\cdot}001 \times 500) \\ &= \exp(-0{\cdot}5) \\ &= 0{\cdot}61 \text{ (using tables of the exponential function)} \end{aligned}$$

We can also calculate the fraction of parts originally put on test, which are failing per hour at this instant when the time is 500 hours, since

$$\begin{aligned} f(t) &= \lambda \exp(-\lambda t) \\ &= 0{\cdot}001 \times 0{\cdot}61 \\ &= 0{\cdot}000\,61 \text{ failures per hour} \end{aligned}$$

It is interesting to plot $R(t)$ and $f(t)$ against time, to ascertain their general shape, and this is shown in Figs. 3.2(a) and 3.2(b).

In Fig. 3.2(a) notice that:

1 λ being constant, plots as a horizontal straight line.

2 Since $f(t) = \lambda \times R(t)$

a At the start of the test where $R(t_0) = 1{\cdot}0$, we have $f(t_0) = \lambda$.

b Because λ is constant, the curves for $f(t)$ and $R(t)$ are identical in shape. Only the scale on the left hand side is different.

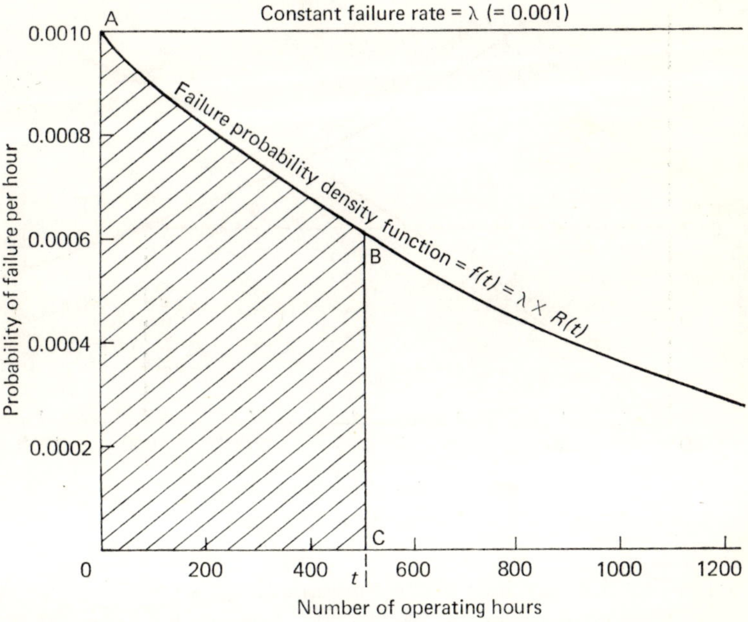

FIG 3.2(a). Failure rate and failure probability density function against time

In Fig. 3.2(b) notice that:

In addition to the reliability $R(t)$, we have plotted the unreliability $F(t)$. Thus:

$R(t) =$ Probability an item will still be working at time t,
$F(t) =$ Probability an item will not be working at time t.

Now at any given instant, an item must be either working or not working, since there is nothing else it can do.

$$\text{Therefore: } R(t)+F(t) = 1{\cdot}00 \qquad (3.2d)$$

Statisticians would say that $R(t)$ and $F(t)$ are *mutually exclusive*, meaning that every item must be one or the other. This can be

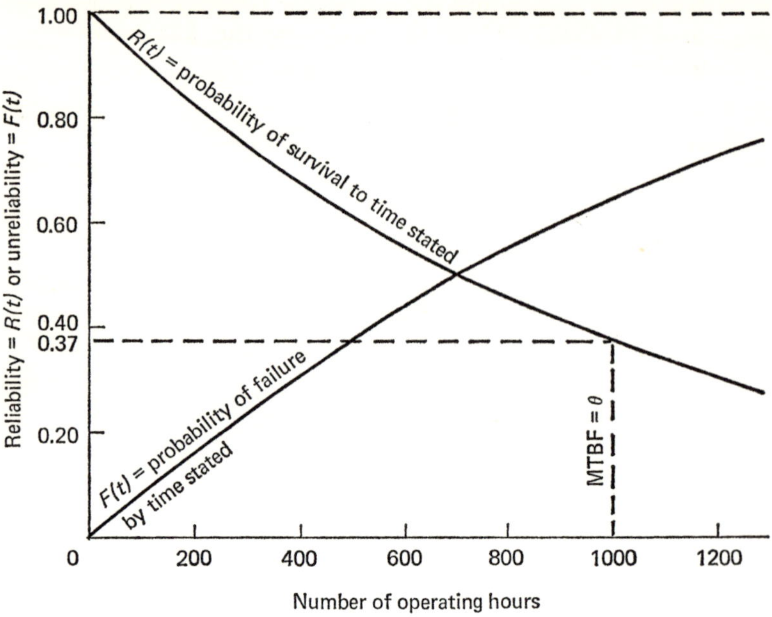

FIG 3.2(b). Reliability and unreliability against time

seen in Fig. 3.2(b), where $F(t)$ is the same shape as $R(t)$, but upside down. For any instant we choose, the sum of the values of the curves for $R(t)$ and $F(t)$ is 1·00.

3.3. properties associated with a constant failure rate

When testing for reliability, it is often convenient to express the duration of the test as so many times the MTBF. Thus, suppose we operate some parts for a time equal to their MTBF. From equation (3.2b):

$$R(t) = \exp(-\lambda t)$$

We insert $t = \theta = 1/\lambda$ from equation (2.9a). Therefore:

$$\begin{aligned} R(t) &= \exp(-\lambda \times 1/\lambda) \\ &= \exp(-1) \\ &= 0{\cdot}36788 \text{ approx.} \end{aligned} \tag{3.3a}$$

Therefore, any item with a constant failure rate which is

operated for a period equal to its mean time between failures will have a reliability of 0·36788 or roughly 37 per cent. Of course, if an item is very reliable, its MTBF will be correspondingly long, so it will have to be operated for a long time before its reliability falls to 37 per cent. Since a reliability of 0·37 is usually completely unacceptable, it follows that where the failure rate is constant the useful working life, up to B in Fig. 2.1(a), must be much less than the MTBF.

The above calculation can be made for any fraction or multiple of the MTBF, and the following are some of the possibilities.

Time in units of MTBF	*Reliability*
θ	0·367 88
0·1θ	0·904 84
0·01θ	0·990 05
0·001θ	0·999 00
0·000 1θ	0·999 90
0·000 01θ	0·999 99

Sometimes it is convenient to draw a reliability curve with the time scale in units of MTBF instead of in hours. When so drawn, it is called a *standardised curve*.

A PRACTICAL CHECK FOR CONSTANT FAILURE RATE

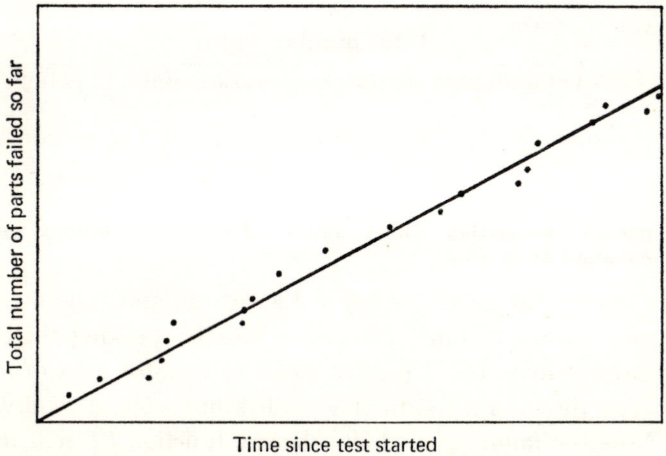

FIG 3.3(a). Cumulative number of failures against time

It is useful to have a simple check to see whether any particular failure rate is constant. To do this we put a number of items on test and :

1 Note the time of each failure.
2 Replace it with a good item, so that the number on test remains constant.

We then plot a cumulative curve, as shown in Fig. 3.3(a). If the failure rate is constant, then a straight line will result, which goes through the origin.

1 If the line is straight, but cuts the vertical axis above the origin, one possible explanation is that extra early failures are occurring at the start.
2 If figures are available in which the items were not replaced as they failed this graph can still be plotted, but horizontally we must now plot the total hours on test of all the items added together.
3 To estimate the failure rate we use :

$$= \frac{\text{Total number failed}}{\text{Total test hours of all items added together}}$$

Where the items are replaced as they fail, this becomes :

$$= \frac{\text{Total number failed}}{(\text{Total number on test}) \times (\text{Duration of test in hours})}$$

We shall discuss this again in sub-section 3 of section 20.6.

3.4. general properties which apply whether the failure rate is constant or varying

Refer now to the curve for $f(t)$ in Fig. 3.2(a), and consider some arbitrary instant in time, marked t. Strictly speaking the curve refers to one item, but it is often easier to visualise a lot of items, so that at the instant t some have failed and some are still working. Now the failure probability density function $f(t)$ tells us the fraction of those which were good at the start, which fail during each instant of time. If we sum all these values together from the start of the test, to the instant t we are considering, it will give us the total fraction which have failed so far.

Therefore area OABC = Total fraction failed from $t = 0$ up to time t
$= F(t)$

At the instant t, the fraction which survives represents the reliability at that moment, but if we continue to run the test long enough, all the items will eventually fail.

Therefore area under the curve from BC to infinity is the reliability at time t.

We can write these results mathematically as follows.

$$\int_0^t f(t)\mathrm{d}t = F(t) \tag{3.4a}$$

$$\int_t^\infty f(t)\mathrm{d}t = R(t) \tag{3.4b}$$

and of course

$$R(t)+F(t) = 1{\cdot}00$$

This result also follows from equation (1.7b).

3.5. wear out failures

Parts do not always exhibit a clear cut wear out failure pattern, but where they do, we may find that the failure probability density function is roughly bell shaped, as shown in Fig. 3.5(a). In practice, of course, random failures which have been occurring during the useful working life will not stop because wear out has commenced, although as the number surviving drops rapidly the rate at which random failures occur will also fall.

Sometimes the bell shape approximates to a Normal frequency distribution as shown. This is a theoretical curve whose equation can be deduced from first principles, if two assumptions are made:

1 That all the variables which affect the moment of wear out are very small.

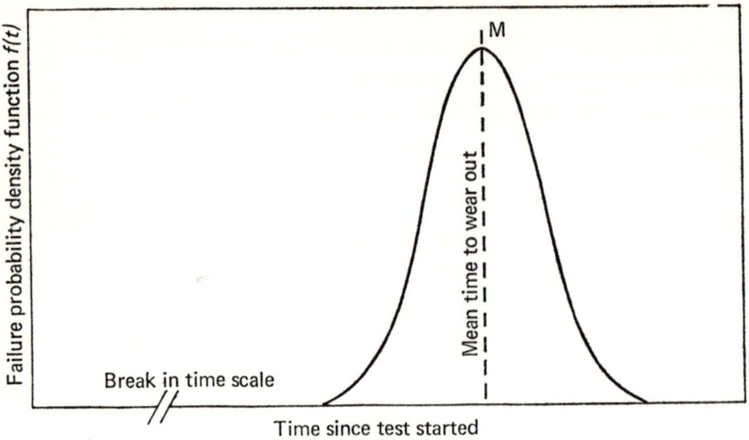

FIG 3.5(a). Typical wear out failure pattern

2 Each variable has an equal probability of being either positive or negative with respect to the mean time of wear out *M*. This means that each variable is just as likely to make the part last a little longer, as to make it fail a little sooner than the mean time.

The variables concerned may be small variations in the materials used, method of manufacture, conditions of use or test.

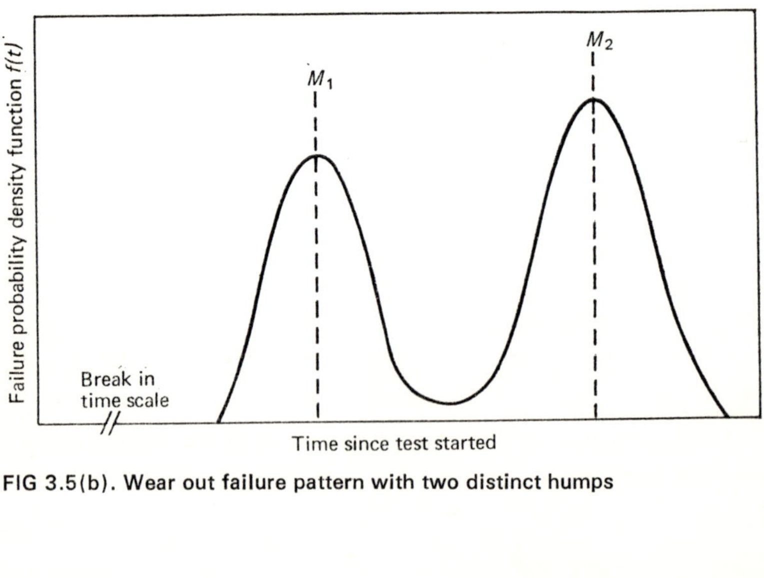

FIG 3.5(b). Wear out failure pattern with two distinct humps

The important thing is that if the wear out pattern is obviously not Normal, then one or both of these conditions have been infringed, and it is worthwhile to try to find out why.

Thus suppose that the wear out pattern has two distinct humps as shown in Fig. 3.5(b). Clearly some of the parts have a much shorter mean life M_1, compared with those associated with hump M_2. We shall first look to see whether both groups have failed in the same way. If we find that failures in the M_1 group are caused by a different fault from those in the M_2 group, we shall concentrate first on the M_1 group, because these occur earlier. If we can eliminate them, we shall have a roughly Normal distribution centred on M_2.

Even if the fault is the same, there must be some difference in the way it occurs. We shall investigate whether M_1 and M_2 are each associated with different:

Test rigs,
Batches of material,
Operators,
Machines, etc.

If the difference proves to be associated with the test conditions, there may be no real difference between the parts in M_1 and M_2. Unless we know this, however, a lot of time may be wasted by production engineers trying to identify a difference in the conditions of manufacture. No method of test is ever absolutely accurate, and often the errors are much greater than is realised.

Figs 3.5(c) and 3.5(d) show wear out failure distribution curves with long tails. In practice there is often not enough test information to plot a really smooth curve, but a scatter of odd points might indicate the presence of a tail in one direction.

Where the tail effectively shortens the life span, as in Fig. 3.5(c), the production engineers will have to identify the particular stage of manufacture which causes it, and then try to eliminate the unwanted variables. At first sight, the pattern shown in Fig. 3.5(d) where some parts last appreciably longer than the rest, need cause us no concern. Up to a point this may be right, but the fact remains that if we can make perhaps 20 per cent of our

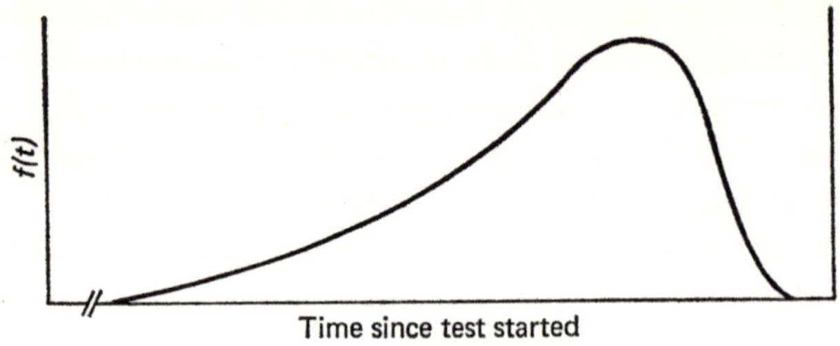

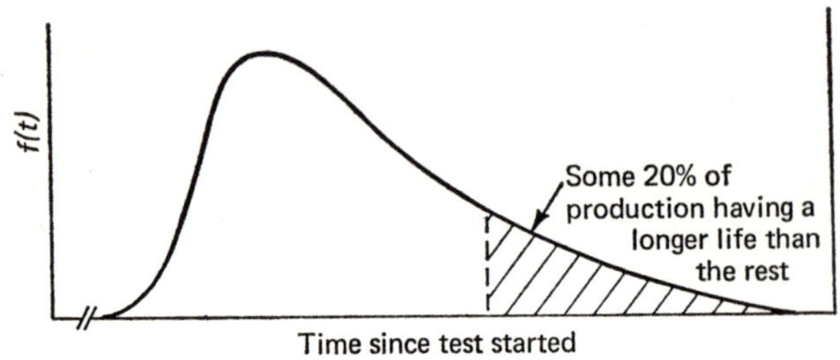

FIG 3.5(c) and 3.5(d). Wear out failure patterns with a long tail

production better than the rest, why cannot we do it for the other 80 per cent as well. Thus again, the production engineers must identify the stage of manufacture at which the difference occurs, but this time when they find it they will try to make the bulk of the production comply with the 20 per cent.

3.6. the useful life of parts

Even between wear out curves which are Normal in shape, there may be considerable differences. Suppose we are testing apparently identical parts from each of two manufacturers A and B, and we obtain the two curves shown in Fig 3.6(a). Both have the same mean time to wear out M, but B has a much greater variability than A. The result is that the first wear out failures from B occur approximately at time W_B, whereas parts from A do not begin to wear out until W_A. Preferably therefore we shall buy from manufacturer A rather than from B.

Now mean time to wear out is exactly what it implies. At time M, approximately half the parts will still be working and half will have already failed, and this is equivalent to a reliability of only 0·50 even if we neglect random and other failures. If we wish to avoid wear out failures we must replace parts from manufacturer B at time W_B, whereas those from manufacturer A need not be replaced until time W_A. To be able to do this, however, we must have some idea of when W_A and W_B occur, which means that we want to know how far curves A and B spread on each side of the mean value M. This is measured by

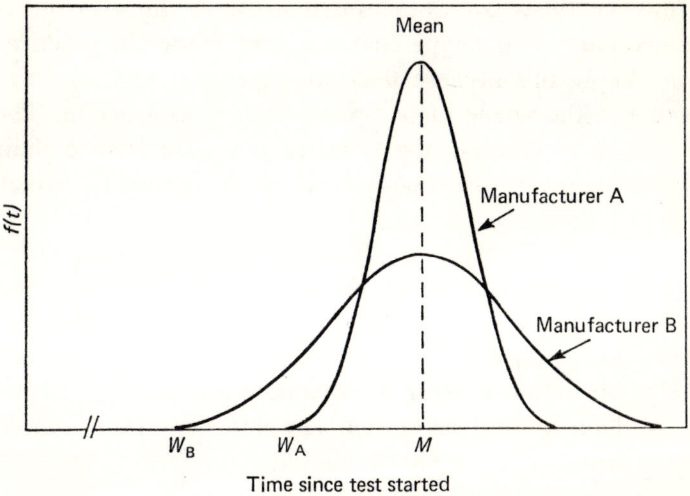

FIG 3.6(a). Wear out failure curves with the same mean time to wear out, but different variability

the *standard deviation*, for which the Greek letter σ is used. Standard deviation is defined by the expression :

$$\sigma = \sqrt{\left[\frac{\Sigma(t_f - M)^2}{N}\right]} \qquad (3.6a)$$

Where

M = Mean time to wear out failure of all the parts of that type ever made.

t_f = The actual time to wear out failure of each part taken in turn.

N = The total number of parts considered in M and t_f.

If we know the value of the standard deviation, then it is a fundamental property of a Normal distribution that, for practical purposes, wear out failures will begin to occur at a time equal to $(M-3\sigma)$. Thus if the mean time to wear out is 3000 hours and the standard deviation is 100 hours, then wear out failures will begin at about time W where:

$$W = 3000-(3\times 100)$$
$$= 2700 \text{ hours}$$

However, the above assumes that we know the wear out time of every part of that type that was ever made. In practice we do not know this because tests are expensive, and anyway we cannot test the whole of our production to destruction. Therefore, we must estimate the standard deviation from a sample, and possibly a very small sample. To do this we modify equation (3.6a) as follows:

$$\text{Estimate of standard deviation} = \hat{\sigma} = \sqrt{\left[\frac{\Sigma(t_f-M)^2}{n-1}\right]} \qquad (3.6b)$$

where

M = Mean time to wear out of sample.
t_f = Time to wear out of each part in the sample.
n = Number of parts in the sample.

Thus, suppose we have only 5 observations of wear out failure as follows: 5700, 6000, 6100, 6200, 6500 hours.

1 We first find the mean,

$$M = \frac{30{,}500}{5} = 6100 \text{ hours}$$

2 We now take 6100 from each observation in turn, and square what is left,
$-400^2, -100^2, 0^2, +100^2, 400^2$
$= 160{,}000+10{,}000+0+10{,}000+160{,}000$

3 Add all these squares together,
$= 340{,}000$

4 Divide by $(n-1) = (5-1) = 4$
$= 85{,}000$

5 Take the square root,
$= 291{\cdot}5$ hours

The true standard deviation may not be exactly 291·5 hours, but this is the best we can do with the limited information available to us. Therefore, we can now estimate that:

$$\text{Wear out will start at about} = (M - 3\sigma) \qquad (3.6c)$$
$$= 6100 - (3 \times 291{\cdot}5)$$
$$= 5225{\cdot}5 \text{ hours}$$

As our estimate is unlikely to be accurate, we might round this off to 5200 hours, or even 5000 hours.

3.7. plotting a Normal frequency distribution curve

Suppose for simplicity that the mean time to wear out M is 1000 hours and that the standard deviation is 100 hours. Notice that

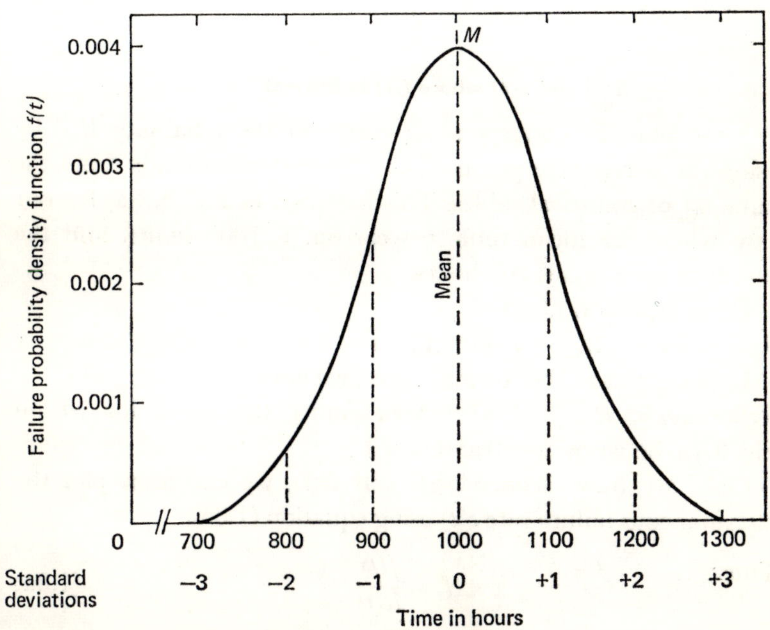

FIG 3.7(a). The Normal frequency distribution curve

the standard deviation is always in the same units as the mean, in this case in hours. From this information, we can easily plot the corresponding Normal distribution curve, using a set of statistical tables, giving the ordinates of the Normal distribution. [*Statistical Tables for Science, Engineering and Management* by J. Murdoch and J. A. Barnes, published by Macmillan.] Such tables give the horizontal scale in units of standard deviation, but this presents no difficulty. Starting with a time scale in hours as in Fig. 3.7(a), we mark the position of the mean M and then on each side of M we scale off in units of the standard deviation. We now plot our curve using the values given in the tables.

Finally we must make a small correction. The tables are in units of standard deviation, and so our vertical scale $f(t)$ will show the probability of failure in a time unit of one standard deviation. As we shall want it per hour, we must divide the vertical scale by the number of hours in a standard deviation. Thus, in Fig. 3.7(a), the ordinates in the table have been divided by 100 in order to plot $f(t)$.

3.8. Plotting *R(t)* and *z(t)* where *f(t)* is Normal

We can plot the corresponding curve for the reliability $R(t)$ by using the percentage points of the Normal distribution from the same set of statistical tables. This is shown in Fig. 3.8(a) for our case where the mean time to wear out is 1000 hours, and the standard deviation is 100 hours.

From equation (3.2d):

Reliability $= R(t) = 1{\cdot}00-$(Fraction failed so far).

Relating this to the Normal curve, we have:

Reliability $= R(t) = 1{\cdot}00-$(Area under the curve for $f(t)$ in Fig. 3.7a, between time 0 and time t.)

Once we have deduced $f(t)$ and $R(t)$, we can soon plot the instantaneous failure rate $z(t)$ using equation (1.8a).

Thus:

$$z(t)=\frac{f(t)}{R(t)} \qquad (1.8a)$$

This is shown in Fig. 3.8(b).

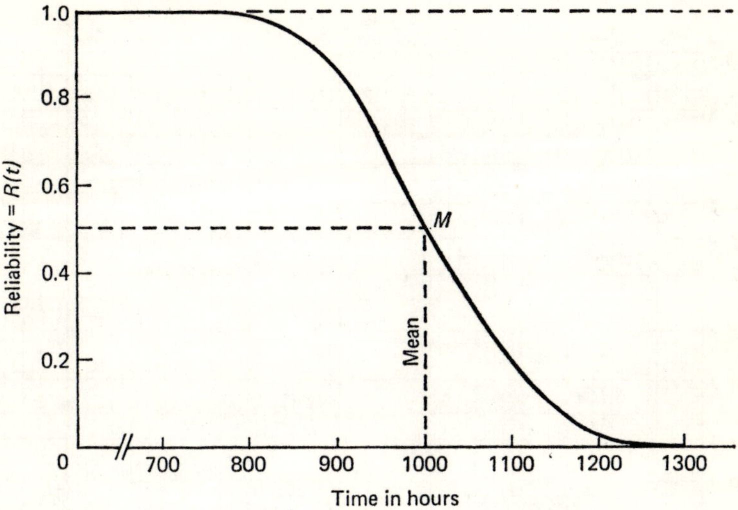

FIG 3.8(a). Reliability against time in operation for a Normal frequency distribution

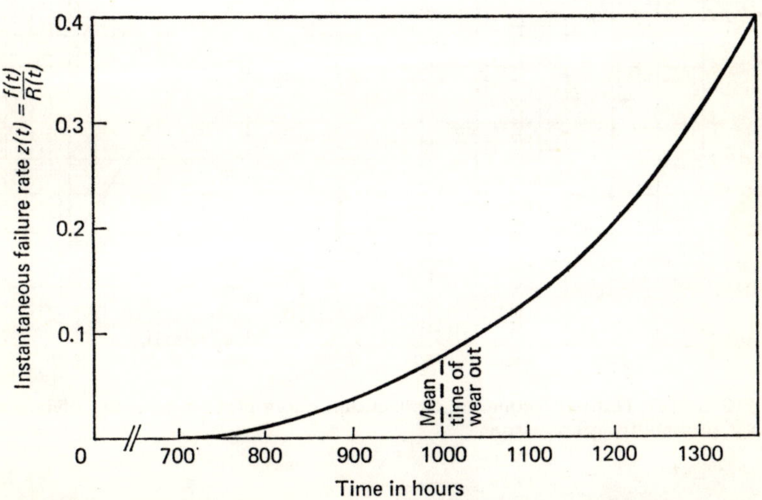

FIG 3.8(b). Instantaneous failure rate against time in operation for a Normal frequency distribution

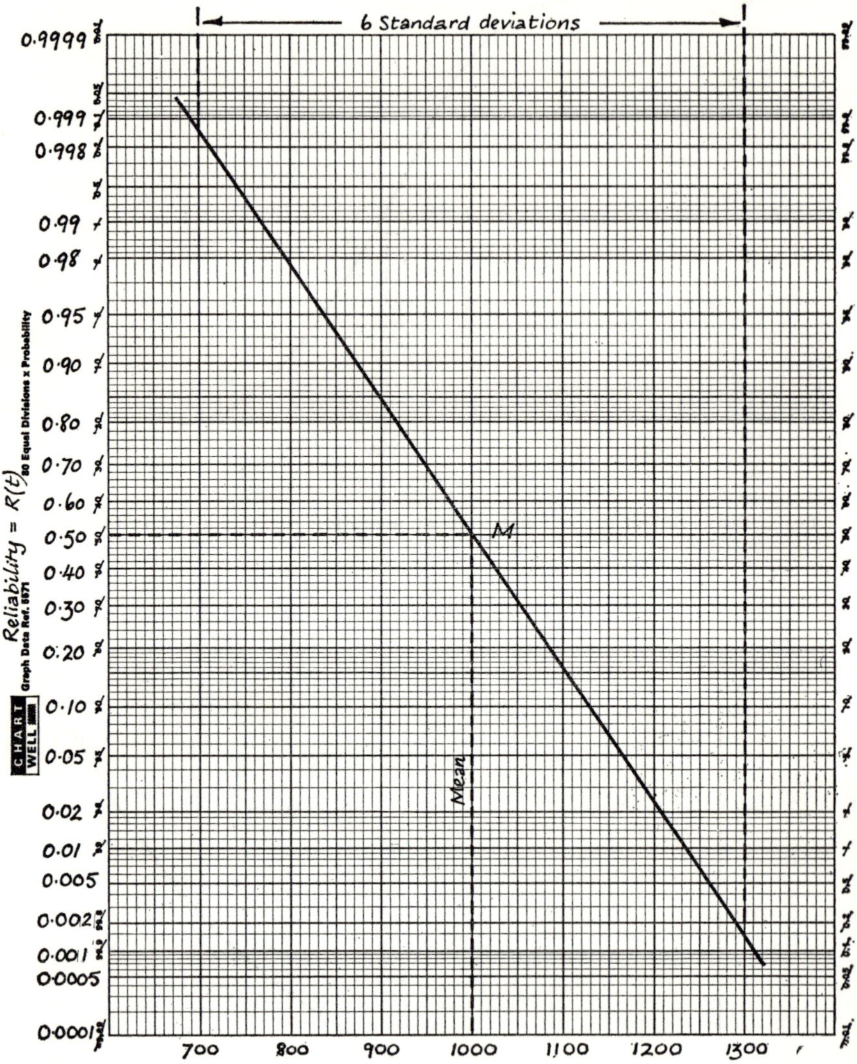

FIG 3.9(a). Normal frequency distribution curve plotted on equal divisions vs. probability graph paper

Notice that the instantaneous failure rate $z(t)$ increases with time at an ever greater rate. This must be so because $f(t)$ in Fig. 3.7(a) is symmetrical, so that the fraction of the original parts failing at, say, time $(M-1\sigma)$ is the same as at time $(M+1\sigma)$. However by time $(M+1\sigma)$, the majority of parts which were working at $(M-1\sigma)$ have failed. Hence $z(t)$, the probability of failure in unit time, as a fraction of those surviving is much greater.

3.9 a practical check for a Normal frequency distribution curve

An easy way to check whether any observed results form a Normal frequency distribution curve is to plot the reliability curve on equal divisions vs. probability graph paper, which is available in the Chartwell range.

Basically we draw a graph exactly as Fig. 3.8(a). However, the vertical probability scale is not linearly spaced, but has been 'stretched out' at the top and bottom. The degree of stretching is such that if the distribution is Normal, the curve in Fig. 3.8(a) will plot as a straight line. Fig. 3.9(a) shows this for the example we have been considering, where the mean is 1000 hours and the standard deviation is 100 hours.

The steeper the line, the smaller the standard deviation. Indeed we can easily estimate the standard deviation from Fig. 3.9(a) as follows. We note the number of hours between reliability or probability 0·001 35 and 0·998 65, and divide this by 6. Thus, in our example there are 600 hours between these limits, so dividing by 6 gives a standard deviation of 100 hours, which we already know it to be. The reason for these apparently curious figures is that in theory there is only a probability of 0·001 35 that a part will fail before time $(M-3\sigma)$, and the same that it will last after $(M+3\sigma)$.

3.10. the search for a more flexible distribution

We have now examined two particular distributions in some detail :

1 The exponential distribution, which applies where the failure rate is constant.

2 The Normal distribution which describes the pattern of wear out failures provided the conditions set out in section 3.5 are met.

Clearly these are two special though important cases, and there are likely to be many failure patterns in practice which do not exactly fit either. Reliability engineers therefore looked for a frequency distribution which would be more generally applicable, and which would reduce to either the exponential or Normal when required. A number of distributions have been tried, including the gamma, but the distribution which has been found to be most generally useful is the Weibull.

If we were asked to draw any given frequency distribution curve, there would be three things we should want to know.

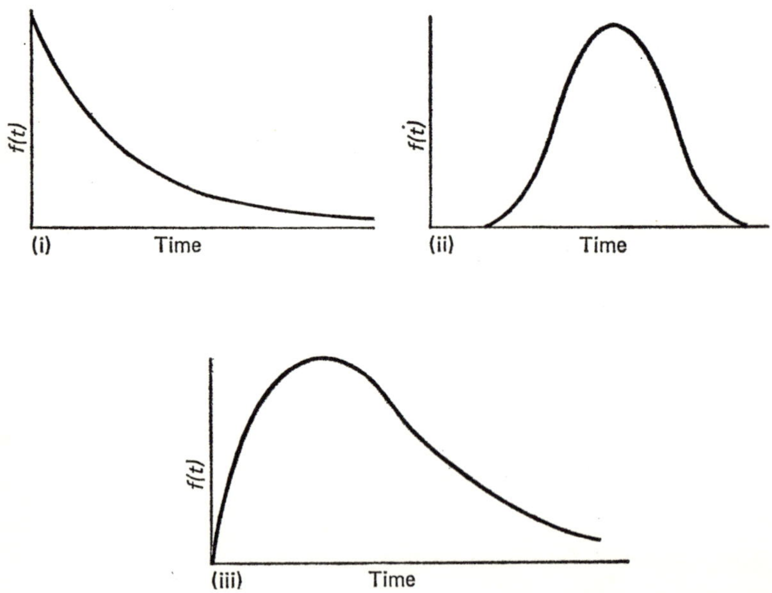

FIG 3.10(a). Examples of frequency distribution curves with different shapes

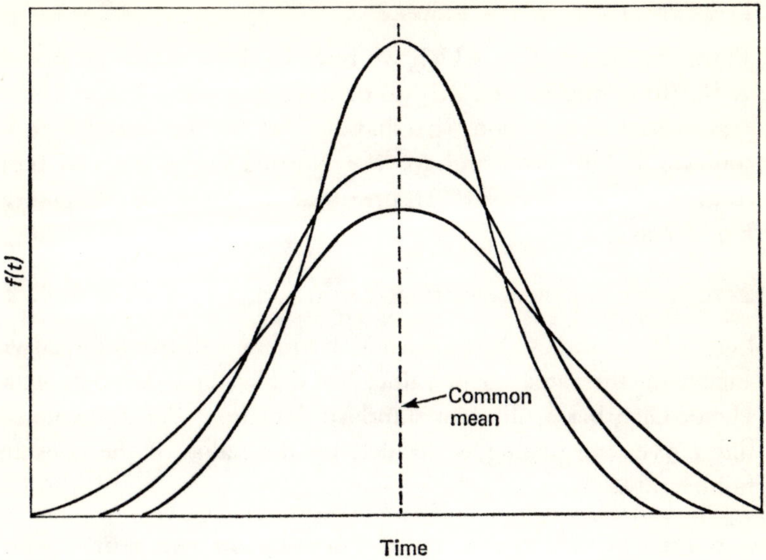

FIG 3.10(b). Examples of Normal frequency distribution curves drawn to different scales

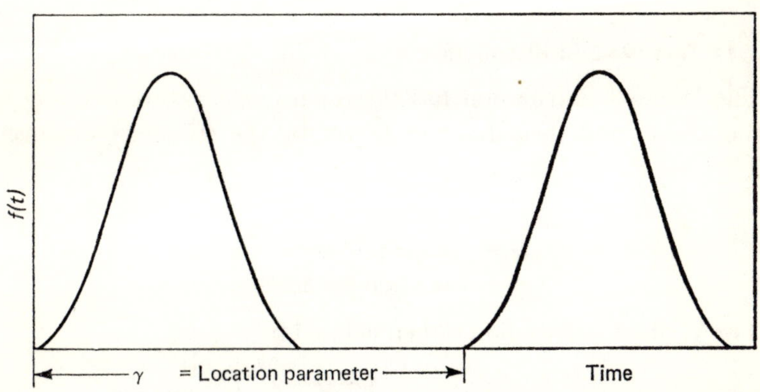

FIG 3.10(c). Two otherwise identical Normal frequency distribution curves, in which the location is different

1 WHAT IS TO BE ITS SHAPE?

Thus, the distribution which we have to draw might be shaped as (i), (ii) or (iii) in Fig. 3.10(a) or have any other shape. In the exponential distribution, the shape is set by the assumption of constant failure rate, and for the Normal curve the two basic conditions we discussed ensure that it must be shaped as Fig. 3.7(a).

2 TO WHAT VERTICAL AND HORIZONTAL SCALE IS IT TO BE DRAWN?

Fig. 3.10(b) shows three Normal frequency distribution curves each with the same mean value, but drawn to a different scale. Hence each has a different standard deviation. For an exponential curve, the scaling is decided by the value of the constant failure rate.

3 WHERE IS THE CURVE TO BE LOCATED ON THE TIME SCALE?

The two frequency distribution curves in Fig. 3.10(c) are identical in shape, but located differently on the time scale. The position of a Normal curve on the time scale is decided by the value of the mean M, whereas an exponential curve always starts at time nought.

3.11. the Weibull distribution

The Weibull distribution has three parameters corresponding to the above, and these have to be set by the reliability engineer. They are:

$$\alpha = \text{Scale parameter.}$$
$$\beta = \text{Shape parameter.}$$
$$\gamma = \text{Location parameter.}$$

The Weibull distribution is then defined by:

$$R(t) = \exp\left[\frac{-(t-\gamma)^{\beta}}{\alpha}\right] \tag{3.11a}$$

There is nothing all embracing about the Weibull distribution, but it has been found to fit a very large number of practical cases.

Its construction is however quite logical, if we compare it with equation (3.2b) for the exponential case. By subtracting the constant γ from t, we ensure that the curve will not start until a time γ after time nought. In many practical cases we want the curve to start at time nought, so if we put $\gamma = 0$, we get

$$R(t) = \exp\left[\frac{-t^{\beta}}{\alpha}\right] \tag{3.11b}$$

and we can deduce the other relations, since from equation (3.2a)

$$f(t) = \frac{-\mathrm{d}R(t)}{\mathrm{d}t} = \frac{\beta t^{(\beta-1)}}{\alpha} \exp\left[\frac{-t^{\beta}}{\alpha}\right] \tag{3.11c}$$

and from equation (1.8a)

$$z(t) = \frac{f(t)}{R(t)} = \frac{\beta t^{(\beta-1)}}{\alpha} \tag{3.11d}$$

Now β is the shape parameter, and if we set $\beta = 1$ the above equations reduce exactly to the exponential distribution. Thus:

$$R(t) = \exp\left[\frac{-t}{\alpha}\right] \tag{3.11e}$$

$$f(t) = \frac{1}{\alpha} \exp\left[\frac{-t}{\alpha}\right] \tag{3.11f}$$

$$z(t) = \frac{1}{\alpha} \tag{3.11g}$$

The above equations are identical to the exponential distribution if we let

$$\text{MTBF} = \theta = \alpha$$

Thus, for the exponential case only, the constant α denotes the mean time between failures.

The Weibull distribution will not reduce exactly to a Normal curve, but $f(t)$ approximates very closely to it for values of β around 3·4.

This is illustrated in Fig. 3.11(a), where $f(t)$ is plotted for values of $\beta = 0{\cdot}5$, 1, 2, 3 and 4, while keeping $\alpha = 1$ and $\gamma = 0$. Thus the curve for $\beta = 1$ corresponds to Fig. 3.2(a), and for $\beta = 3$ or 4 it takes on a near Normal shape.

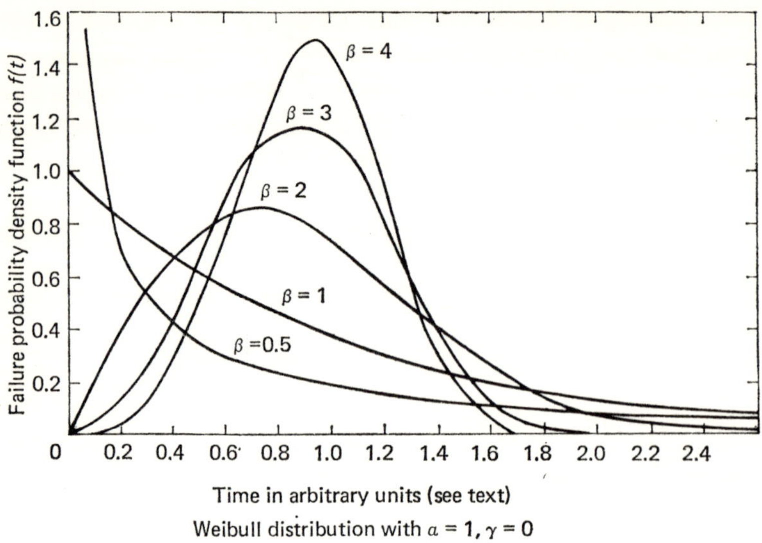

FIG 3.11(a). Typical Weibull frequency distribution curves for failure probability density function

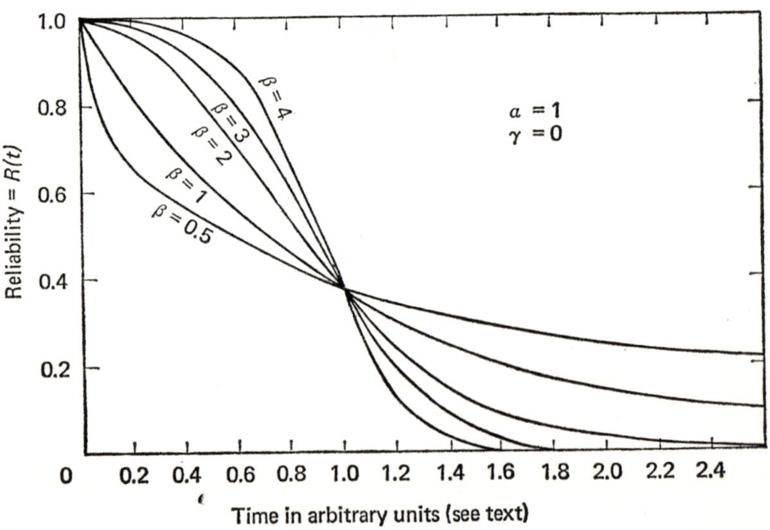

FIG 3.11(b). Typical Weibull frequency distribution curves for the reliability function

By putting $\alpha = 1$, the time scale which is plotted horizontally looks curious because the whole of the useful part of the curves occurs before time $t = 2{\cdot}4$ units, and in almost any practical case we should be very disappointed if most of the items had failed in a mere 2·4 hours. The time scale is however arbitrary, in that there is nothing to say that one of our units must equal 1 hour. We can make it what we like, and in many problems it may be convenient to let one arbitrary unit equal 1000 hours or even more. The only point to remember is that the constant α will be in the unit we choose.

Fig. 3.11(b) shows the corresponding curves for reliability $R(t)$, and again the time scale is expressed in arbitrary units. Notice that all the curves meet at the point where the time $t = 1{\cdot}0$. This follows from equation (3·11b) because, if $t = 1$, it does not matter what value we assign to β since t^{β} will always equal one. The reliability then depends only upon the value given to α.

Fig. 3.11(c) shows the corresponding instantaneous failure rate curves $z(t)$ plotted from equation (3.11d). For $\beta = 1$, $z(t)$ is con-

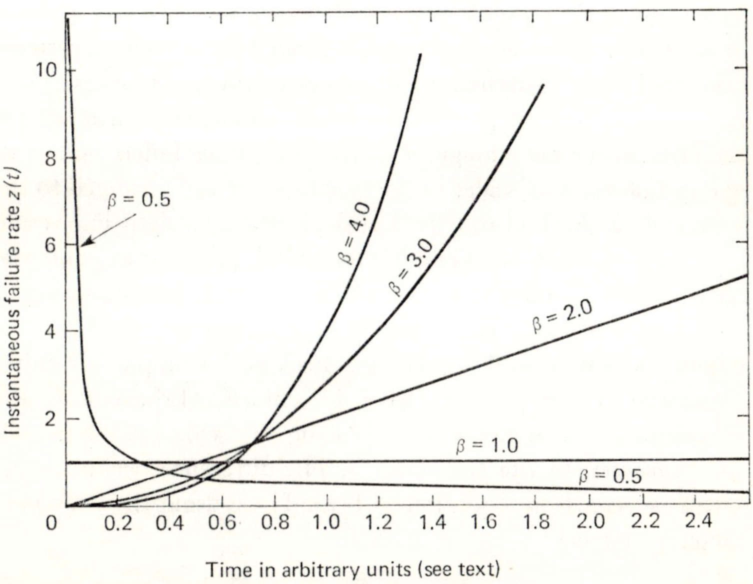

FIG 3.11(c). Typical Weibull frequency distribution curves for instantaneous failure rate

stant as we should expect, and as β gets larger so the curve for $z(t)$ becomes steeper.

3.12. using the Weibull distribution to describe the bath tub curve

The bath tub curve which we discussed in sections 2.1 and 2.4 can be regarded as a series of Weibull frequency distribution curves where:

β is less than 1 during the early failure period.
$\beta = 1$ during the constant failure rate period.
β is greater than 1 during the wear out period.

The other constants must be adjusted to suit. Thus, γ is only likely to be zero for the early failure period, and may take quite a high value for the wear out period.

3.13. estimation of the Weibull parameters

It is quite easy to check graphically to see whether any given set of reliability observations can be described by a Weibull distribution, and then to estimate the values of the parameters. To do this we use Weibull probability graph paper, and on it we plot the cumulative percentage of items which have failed against the age at failure. The scales of the graph paper are arranged so that whenever a Weibull distribution is plotted a straight line results.

Several designs of Weibull probability paper exist, and they differ chiefly in the provisions which are made for estimating the values of the parameters. Fig. 3.13(a) illustrates the Chartwell design, which is sold by H. W. Peel and Company Limited, Chartwell House, Jeymer Drive, Greenford, Middlesex. In this, simplification is achieved by replacing the scale parameter α by the *characteristic life parameter* η. The difference between α and η can be seen from equation (3.11a). Thus, using the scale parameter α we have:

$$R(t) = \exp\left[\frac{-(t-\gamma)^{\beta}}{\alpha}\right] \tag{3.11a}$$

Compare this with the following, which uses the characteristic life parameter η:

$$R(t) = \exp\left[-\left(\frac{t-\gamma}{\eta-\gamma}\right)^{\beta}\right] \qquad (3.13a)$$

When $\gamma = 0$ this reduces to:

$$R(t) = \exp[-(t/\eta)^{\beta}] \qquad (3.13b)$$

tion point' which is printed in the top left-hand corner of the paper and draw a perpendicular on to the Weibull line. On its way it will cut the β scale at the best estimate of β.

1 In the examples the location parameter $= \gamma = 0$. If it had had a positive or negative value, the Weibull plot would have been curved, and the example in section 20.11 shows how it would have been estimated.

2 To estimate the shape parameter β, we start from the 'estimation point' which is printed in the top left-hand corner of the paper and draw a perpendicular on to the Weibull line. On its way it will cut the β scale at the best estimate of β.

3 The characteristic life parameter η is obtained by reading the age on the horizontal scale at which 63·2 per cent of the items have failed. The justification for this comes from equation (3.13b), where if we insert $t = \eta$ we get:

$$R(t) = \exp[-(\eta/\eta)^{\beta}]$$
$$= \exp(-1) = 36{\cdot}8 \text{ per cent}$$

Therefore, the total percentage which have failed is:

$$F(t) = 1 - R(t) = 63{\cdot}2 \text{ per cent}$$

Hence, $F(t)$ always has a value of 63·2 per cent for a time equal to the value of η, and so to estimate η we merely read off the time for which the cumulative percentage of failures is 63·2 per cent.

4 The scale marked $P\mu$ allows us to use the perpendicular from the estimation point on to the Weibull line, as described in 2 above,

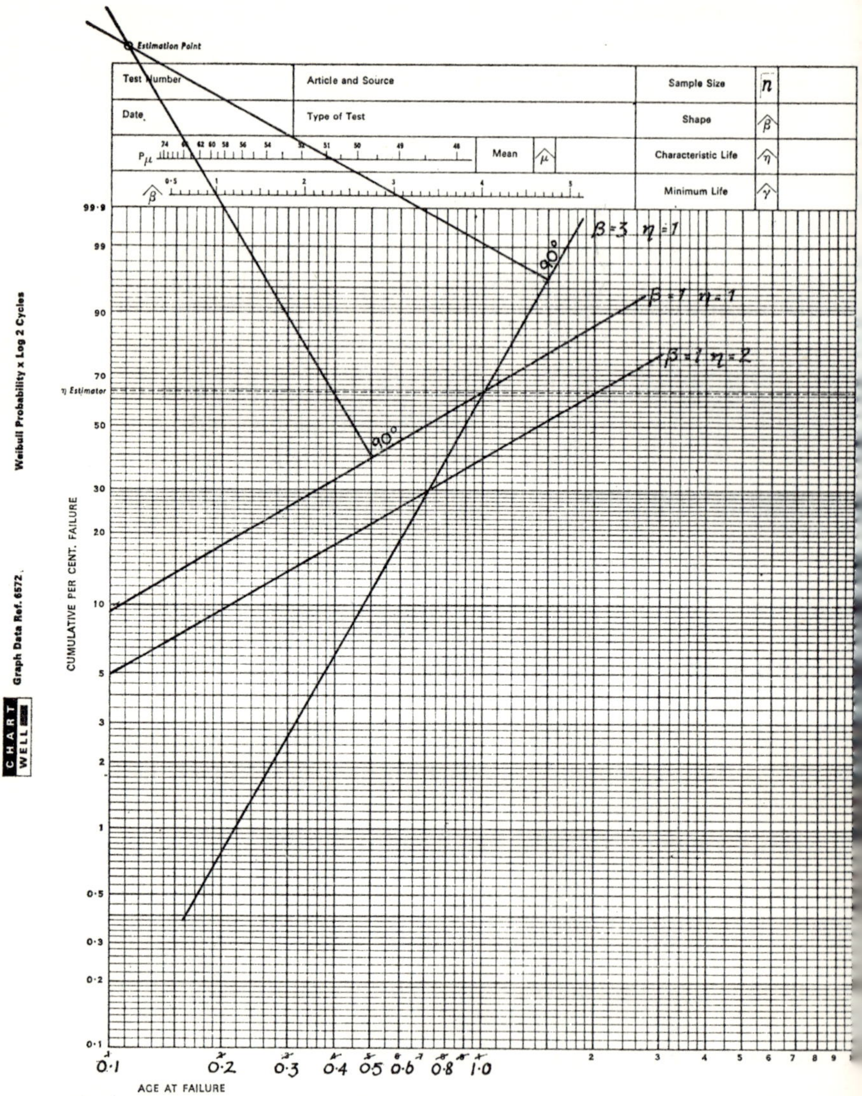

FIG 3.13(a). Examples of the use of Weibull probability paper

to estimate the percentage of items which have failed by a time equal to the mean time to failure. For example, if we use the line where $\beta = 1$, we find that $P\mu = 63{\cdot}2$ per cent in accordance with equation (3.3a), whereas for $\beta = 3$, $P\mu$ is about 51 per cent, suggesting a nearly normal distribution.

5 If this value of $P\mu$ is inserted in the vertical axis of the graph we can read off the corresponding mean time to failure.

CHAPTER 4 COST, PERFORMANCE AND OTHER RELATED FACTORS

4.1. factors related to reliability

So far we have considered reliability in isolation, as though nothing else mattered, whereas it is really only one of many important factors, as the following examples illustrate.

1 COST

There is always the risk that in our efforts to achieve a more reliable product, we may price ourselves out of the market.

2 TIME ALLOWED FOR DELIVERY

If our deliveries are too long, our customer may buy from a competitor. However, the time allowed for delivery determines the time available for design, development, manufacture and test, so if the delivery time is impossibly short both the quality and reliability of our products may suffer.

3 QUALITY AND PERFORMANCE

We must ensure that we supply to satisfy our customers. If our quality and performance are too low, our customers may not buy; if it is unnecessarily high, our prices may also be too high.

4 SAFETY

Safety is clearly important, since we cannot tolerate foreseeable danger to people. In a sense one can regard safety as an aspect of reliability, because an equipment which injures a person is clearly not working 'as intended'. Because of its importance, however, it is usually considered separately.

5 WEIGHT AND VOLUME

Today's trend is towards miniaturization, so that both weight and volume are reduced to a minimum. Obviously this is vital to the aircraft industry, but the implications are much wider. A modern computer would be unacceptably large without the intensive work which has been done to make it as small as possible. At the other end of the scale, we have the ordinary transistor radio which is often small enough to slip into the pocket.

6 EXTENSIBILITY

By this we mean the ease with which the equipment can be extended or enlarged. It is sometimes defined as the minimum size of an economic extension. In industry is it usually difficult to predict accurately what future market requirements will be. The production engineer does not want to buy a very large expensive piece of plant which he may never fully use. Yet if he buys a smaller unit initially, it helps considerably to know that it can be extended later if required. Extensibility is in the supplier's interests too, because having installed the small unit, he knows that if the equipment is extended he stands a very good chance of getting the order.

7 ADAPTABILITY

Adaptability is concerned with the addition of more facilities. Thus if we are installing an office telephone system, we may ask whether push button dialling can easily be added later if required. However many facilities, even though they can be added later, are best and cheapest if included in the new equipment. Central heating and double glazing, for example, can be added to a building at any time, but are cheapest if put in when the building is first erected. Similarly, the addition of automatic transmission to a car is usually relatively costly once the car has been completed with a conventional clutch and gear box.

4.2. the design profile

When we want to buy a new piece of equipment, we at first sight have no hesitation about what we want, for example :

1 Absolute reliability.
2 Extreme cheapness.

3 Immediate delivery.
4 Very cheap and easy to maintain.
5 Excellent performance, quality, etc.
6 Absolute safety.
7 Negligible weight and volume, etc.

Unfortunately the above is quite impossible, because these factors are interrelated. The greater our demands on quality, reliability, performance, safety, weight and volume, the greater the cost is likely to be and the longer the delivery time. Therefore, for each individual case we have to balance one against the other. The reliability engineer often refers to this as being able to 'trade off' one factor against another, although it would probably be more appropriate to use the term optimize. We should think not so much of 'robbing Peter to pay Paul', but rather of the need to obtain a combination of all these factors which best satisfies the true needs of that particular customer.

Thus at one end of the scale, a defence contract for a nuclear deterrent may be considered so important that high performance, and near absolute reliability, in a minimum period of time, are essential. If as a result the cost is very high, this may have to be accepted. At the other extreme, it may be decided that a child's toy must be very cheap and very safe, at the expense if necessary of performance and reliability.

Consciously or unconsciously the designer of any piece of equipment must decide on the optimum mix. This may be difficult, and some degree of judgement is nearly always involved. One approach is to draw a design profile, showing the approximate proportion of each factor which is required, and an example of this is shown in Fig. 4.2(a).

Sometimes arbitrary units are allocated. We may, for example, know roughly what it will cost for every month we, as the manufacturer, are late on delivery, or for how much more an equipment will sell if its reliability is increased by a given amount. In this latter case, we can calculate whether the cost of the extra reliability will be recouped in the higher selling price.

Some care is necessary in the use of arbitrary units of this type.

Once values have been allotted, the calculation to determine the optimum mix may be quite straightforward. If the assumed values are inaccurate, however, the results of the calculations are likely to be equally inaccurate.

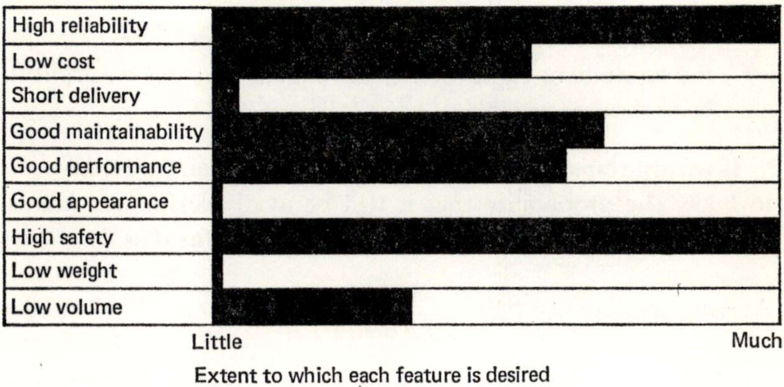

FIG 4.2(a). Example of a design profile

Notice that nothing in the trade-off argument excuses us from attempting to obtain an optimum solution. If the reliability can be increased without putting up the cost or detracting from other factors so much the better. If the higher reliability is not required, then we can investigate the possibilities of the same reliability at lower cost.

4.3. availability

The subject of this book is always referred to as reliability or reliability engineering, but the problem is really wider than this suggests. We shall no doubt do our best to achieve the highest reliability at minimum cost, but we have to recognize that every equipment will break down sooner or later. When this happens, the speed and ease with which it can be repaired become vitally important, and hence we must also consider maintainability which we shall do in detail in Chapters 11 and 16.

For any given equipment, we can express the mean time to

repair (MTTR) in a similar way to the mean time between failures, which we discussed in section 2.9. Thus, we have:

> The *MTBF* telling us how long on average, an equipment operates before it fails, and this we want to be as long as possible.
>
> The *MTTR* telling us how long on average, it takes to put the equipment right after it has failed, and this we want to be as short as possible.

If we are responsible for operating an equipment, we shall want to know the probability that it will be available for use. Hence we use a quantity called *availability*, which is defined as:

$$\text{Average availability} = \frac{MTBF}{MTBF+MTTR} \tag{4.3a}$$

In the case where we want to use the equipment continuously, the average availability expresses the probability it will actually be working. An average availability of 0·80 means that the equipment is working satisfactorily for 80 per cent of the time, and under repair, including waiting for spares, etc. for the remaining 20 per cent.

A quantity called *point availability* is also used. This is the probability that a system or equipment will be available for use at some specified instant of time. Thus the average availability may be 0·80, but if we ask whether the equipment will be available for use at, say, 10.00 hours today, the answer may be no, because it has just broken down and is being repaired. Hence, at that particular moment in time, the point availability will be zero.

The definition of availability given above is typical of the industrial case where, once an expensive piece of equipment has been installed, it will be expected to work as continuously as possible in order to earn its keep. It is less appropriate to the military situation where no one intends that because, say, an expensive missile launching system has been bought, it will continuously fire missiles. For such cases it may be more convenient to use the *mission avail-*

ability. This is given a mathematical definition, but broadly it expresses the probability that a mission of given duration will be successfully completed within a given time.

4.4. utilization factor

Availability as defined by equation (4.3a) represents the maximum utilization which can be obtained from the equipment concerned. In most practical cases there will be occasions when the equipment is not required to operate, because there is no work or mission for it to do or perhaps because the operators are off duty. We can express this as a *utilization factor* if we add another term to the denominator to represent the mean time the equipment is idle between repairs (MTIBR). Thus:

$$\text{Utilization factor} = \frac{MTBF}{MTBF+MTTR+MTIBR} \qquad (4.4a)$$

4.5. system effectiveness

We have now expressed the probability that an equipment or system will work as intended when required to do so, but this may still not be good enough to accomplish the mission. An anti-missile system, when performing exactly as intended, may only be successful in dealing with say 90 per cent of the missiles concerned. If the enemy makes his missiles less vulnerable, that same anti-missile system may then only deal successfully with perhaps 60 per cent of them. This is expressed by the system effectiveness, which is defined as follows.

System effectiveness is the probability a system can successfully meet an operational demand, within a given time, when operated under specified conditions.

Some of the definitions given in this chapter have not been precisely defined, and their meaning varies a little between different reliability engineers. In particular, statisticians tend to use very precise mathematical definitions in order to make it easier to employ these quantities in calculations.

4.6. reliability and maintenance costs

The cost of achieving any desired reliability and the subsequent cost of maintenance are related to each other roughly as shown in Fig. 4.6(a). Consider first the cost of achieving reliability. At A the reliability is very low and, since presumably neither the designers nor the production unit care much about it, the amount spent on reliability is also low. Indeed the reliability could be appreciably improved, without spending much, merely by taking a little more care. Hence costs rise little as reliability is improved. If we go on improving the reliability, however, we gradually reach the situation where all the obvious things have been done, and from now on we shall have to spend increasingly more to achieve a given reliability improvement. If we were unwise enough to demand an impossible reliability of 1·00, costs would sweep away to infinity beyond B.

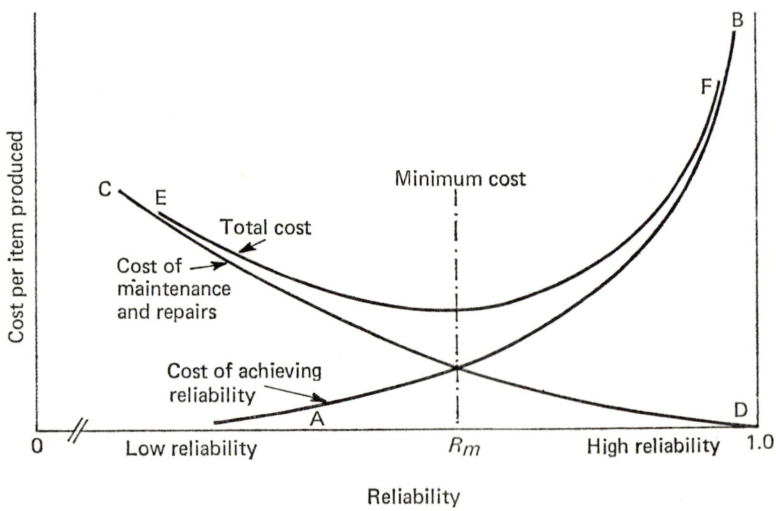

FIG 4.6(a). The relation between reliability and maintenance costs

However, when reliability is low, maintenance costs from all the breakdowns are inevitably high, as shown at C. As the reliability improves so the cost of maintenance falls, until at D, as reliability approaches 1·00, maintenance costs approach zero.

By adding reliability and maintenance costs together we get curve EF, and find that there is a particular reliability R_m for which the overall cost is a minimum.

The above is a useful concept, but it is very difficult to determine money values in an actual case because the practical situation is nearly always more complicated than Fig. 4.6(a) suggests, since:

1 It does not follow that an improvement in reliability must inevitably cost more. Better designs, different materials, better quality control during production, etc. may achieve improved reliability at little or no extra cost. If scrap is reduced at the same time, the overall cost may actually come down. Predicting the cost of achieving any given reliability is nearly always difficult, so that in general we only know the cost accurately afterwards.

2 Maintenance costs are also difficult to estimate. If we know precisely what repair work has to be done each time an equipment fails, then we can predict its cost. However we are unlikely to be able to foresee associated costs such as:

a The value of production lost through breakdowns, including the cost of late deliveries, split batches, idle operators, etc. on the production line.

b The cost of having a piece of military equipment out of action. This probably depends very much on whether it happened to be required for use against the enemy during the time it was under repair.

3 Costs are inextricably mixed up with all the factors discussed in section 4.1. We must define precisely what we mean by reliability costs, or the figures we assign to them will have no meaning.

The fact that accurate reliability costs are so difficult to estimate is of course no excuse for neglecting them. As technology advances, reliability costs continually increase so that in many cases even the roughest of calculations will leave no doubt about the need for intensive control.

4.7. factors affecting reliability and maintenance costs

In principle any project may embrace some or all of the following costs, although they are not related to reliability alone.

1 RESEARCH, DESIGN AND DEVELOPMENT COSTS

a Research into reliability problems for which solutions are not yet known.

b Design costs.

c Cost of building and testing prototypes.

d Cost of modifications to the design, and of further tests until the required reliability is achieved.

2 MANUFACTURING AND INSTALLATION COSTS

a Development and purchase of new manufacturing plant, equipment, tooling, etc.

b Costs of manufacture, including adequate quality control.

c Installation and commissioning costs, when the equipment we have made is installed in our customer's premises.

3 UTILIZATION COSTS

a Day to day running costs, including the costs of routine servicing, repairs, etc.

b Costs of support equipment which is necessary to operate the equipment efficiently.

c Spares, test equipment, etc.

d Training of operators.

4.8. eight basic stages in the achievement of reliability

Achievement of the required reliability can be divided roughly into eight stages, as follows.

1 THE CUSTOMER OR MARKET SPECIFICATION

We must ascertain as accurately as possible precisely what our customers require (see Chapter 5).

2 PREPARE THE DESIGN AND EXPRESS IT AS A MANUFACTURING SPECIFICATION

The designer must specify what has to be made in order to satisfy the requirements of the customers (see Chapters 6, 7, 8, 9, 10 and 11).

3 PROVE THE DESIGN

Wherever the design is a departure from previous practice, we shall need tests on prototypes to show by demonstration that what is proposed will achieve the reliability demanded (see section 12.1 and Chapters 17, 18, 19 and 20).

4 MANUFACTURE TO SPECIFICATION

A high standard of quality control will be necessary to ensure manufacture to specification at minimum cost, in the time required (see Chapter 12).

5 PACKAGING AND TRANSPORT

We must ensure that the equipment is packaged and then transported to the customer's site without incurring any significant damage (see Chapter 13).

6 PURCHASE, STORAGE, INSTALLATION AND COMMISSIONING OF NEW EQUIPMENT

We next look at reliability from the customer's point of view, and consider in Chapter 14 how he decides what to purchase and in Chapter 15 how it is stored, installed and commissioned for use.

7 OPERATION AND MAINTENANCE

The customer must use and maintain the equipment as intended, employing operators with adequate skill and training. If there are any difficulties, the manufacturer should be anxious to help, not merely in the interests of good customer relations but also in order to learn for the future (see Chapters 15 and 16).

8 RELIABILITY MANAGEMENT AND PREDICTION

Finally in Chapters 21 and 22, we consider the overall management of reliability, and how the reliability of a proposed design can be predicted.

PART TWO **DESIGN AND MANUFACTURE FOR RELIABILITY**

CHAPTER 5 CUSTOMER OR MARKET SPECIFICATIONS FOR RELIABILITY

5.1. finding out what the customer wants

It is important to ascertain as accurately as possible :

1 What the customer wants.
2 What he is prepared to pay.
3 When the product must be ready.

Otherwise we may spend a lot of time designing something which later proves to be unacceptable.

Where we have to design and produce for a particular customer, we can obtain this information by direct negotiation. We must take pains to find out exactly what our customer really wants, because this may be different from what he initially requested. In addition to performance and physical features, it is becoming increasingly common for customers to stipulate their reliability requirements.

When we are designing for a general market, information about what is required may not be so easy to obtain and it may come in part from several sources, such as sales staff, sales records, official publications or from a market survey. It is often tempting to design primarily for performance, because this is what a potential customer sees when he contemplates purchase. Unsatisfactory reliability, however, will lead to dissatisfied customers later on, and in the long term may be very damaging.

5.2. customer or market specifications for reliability

Whichever way the information is obtained, it will be necessary to prepare a customer or market specification showing precisely what is required. It is usually best to express this in terms of the service to be performed, or result to be achieved, rather than of the hardware envisaged. Thus suppose a customer asks for 'a reliable electrostatic precipitator, which will ensure that a particular manufacturing process does not discharge to atmosphere more than one milligram per cubic metre of a particular chemical effluent'. This statement restricts the designer, because it excludes other possibilities such as scrubbers, filters and chemical methods. It is better to quote merely the limit of one milligram per cubic metre, and leave the designer to decide how this can best be achieved.

The specification must contain full information about everything which is required, even though in the following sections we shall confine ourselves to factors associated with reliability.

1 SPECIFYING THE RELIABILITY

The required reliability must be expressed in figures, since a statement that an equipment 'must be reliable' has no meaning. What one person means by reliable may be quite different from what someone else means. There are three main ways of expressing reliability in a specification :

a DIRECTLY IN TERMS OF RELIABILITY FOR A SPECIFIED USEFUL LIFE

Because reliability is related to a particular life span, this is not always convenient, and MTBF or failure rate is usually preferred.

b MTBF OR MTTF

This method is common, especially in the electronics industry, where the failure rate is often approximately constant.

c FAILURE RATE

Since the failure rate is directly related to the MTBF, it can be used provided it is reasonably constant.

The reliability of large installations is not necessarily quoted as a single figure. The central unit, for example, may be assigned a higher reliability than some of the ancillary and support equipment. Ideally, the minimum acceptable reliability or MTBF or the maximum failure rate should be quoted, but when the system is of very new design it may be more realistic to set a target reliability, MTBF or failure rate. In deciding what this should be, we must consider the conditions of use, the duty cycle, what the maintenance requirements are likely to be and how long the system can be out of use while maintenance is done.

2 THE ENVIRONMENT AND CONDITIONS OF USE

Next comes the environment in which the equipment will be used. Again this should be quantified, since a statement that 'it will be out in all but the worst weather' means very little. The weather conditions should be stated in terms of temperature and humidity ranges, duration, etc. The terrain may also be important. Thus the designer of a tank which is to be dropped by parachute no doubt hopes it will land on flat firm ground, but it may land in mud or several feet of water. It might conceivably land in a tree and fall to the ground with a crash.

Where a component or unit is to be incorporated into a larger piece of equipment, the designer will need to know where it will be located, and hence the conditions under which it must work. Thus if it is near a source of heat, vibration etc., this will affect its design.

Specifications must also give information about non-operational conditions, and these will include not only those when it is not working but the conditions under which the equipment will be transported to the site and maybe stored until it is required for use. We shall consider these more fully in Chapters 9 and 15.

3 DEMONSTRATING THE RELIABILITY

Our specification must set out how the manufacturer is expected to demonstrate that the required reliability has been achieved. Probably only a limited amount of testing will be possible, and this under controlled test conditions rather than the actual service environment. Therefore we must know what tests are to be done, on how many units and what is or is not to be considered acceptable. It will be necessary to specify the confidence which can be placed in the results, and we shall discuss how to do this in Chapter 19.

4 MISCELLANEOUS INFORMATION

All information which will help the designer should be included, taking care of course not to restrict the designer more than is necessary. Miscellaneous information may include data about:

a The skill of the operators and maintenance staff.
b Maintenance equipment available, etc.

5.3. consulting interested people

We have spoken as though the launching of a project to develop a new piece of equipment only concerns the customer, salesman and designer. There are however others to be consulted whenever a contract is in prospect which will affect them. The factory needs some idea of the work load, and when to expect it. It may also be apparent that some parts of the contract will require work which the factory is not equipped to do. A decision will then have to be made whether to buy new factory equipment or to sub-contract that part of the work. In a few cases it may be more prudent to refuse the contract than to take on work which cannot be successfully completed. The quality manager can help with this because he should have up to date information about the capability of each machine and process. Test facilities must also be considered, otherwise we may discover when the equipment is almost built that we cannot carry out the demonstration tests which are demanded.

5.4. analysing the requirements of a contract

It is unusual for a company to agree to design, develop and make a large system which is completely new, and therefore very different from anything they have previously made. This may happen with single components, but usually a large system is at least partially similar to others they have made in the past and so they have some experience upon which to draw.

An early step is to break the whole contract down into a number of sections, each of which is as self-contained as possible.

Thus the complete system for launching an artificial satellite might be divided into the following :

1 The satellite itself.

2 The launch vehicle.

3 The ground check out equipment.

4 The ground tracking stations.

Launch equipment for an artificial satellite is of course a very large system, but the principle applies equally if it is smaller. Thus we might have a contract to design and build a special piece of automatic moulding equipment for a factory. This might divide conveniently into :

1 The press itself with its dies, etc.

2 A unit which feeds raw plastic material into the press, and removes the product when processing is complete.

3 A control unit which looks after :

a The timing and mechanical sequence of operations.

b The press temperature.

To each of these sections we give a specification which is as complete as possible within itself. In particular, the overall reliability or failure rate is broken down and a target allocated to each section. Thus if our plastic moulding press was permitted 6 failures per 5000 operating hours, we could if we wished allocate 2 failures to each of the three sections. More likely however the allocation would be on the basis of need so that the press itself, which may be a straightforward modification of a standard product, might only be allowed 1 failure, the mechanical feed unit 2 and the control unit, which has some special requirements, 3 failures per 5000 hours. This is considered more fully in Chapter 7.

CHAPTER 6 THE RELIABILITY OF PARTS AND COMPONENTS

6.1. two important rules

Whatever our system it will be made up of parts and components, and since in some cases the failure of one of these may cause the whole system to fail, we must ensure that each is as reliable as possible. Further, the greater the number of parts, the greater the risk of including one which is faulty. Hence we have two basic rules:

1 Use as few parts as possible.
2 Ensure that each part is reliable.

6.2. minimize the number of parts

The old adage that 'what isn't there, can't go wrong', applies with a vengeance to reliability. The modern trend is for systems to become more and more complex. Something is always being added to improve performance, such as make the system work faster, handle heavier loads or do something which it was previously unable to do. Unfortunately, all these additions provide something else to go wrong, so that as performance improves, reliability tends to decrease unless action is taken to prevent this happening.

We can illustrate this with an industrial example by considering the case where batches have to be treated in a succession of chemical baths. In its simplest form, the plant may only consist of a line of tanks, with one operator to immerse batches for the preset times. The method is simple and reliable because there is hardly anything to go wrong. It is also very flexible, since an instruction to the operator is all that is required to change either the preset times or the sequence of bath treatment. It is hardly efficient, however, because output is limited to what one operator can handle.

As the required output grows, the inefficiency becomes more apparent. Merely duplicating the existing method would require a large floor area for the baths, with lots of operators busily lifting batches in and out. For a small increase in output, we might give the operator a hoist, but for a larger increase we would probably mechanize the whole operation. One operator would then feed batches to the equipment, which would immerse them with the required sequence and timing and deliver them to the far end for another operator to remove. Although this would be more efficient, a breakdown would, from time to time, bring the whole production line to a halt. Thus reliability has been reduced, and we may decide to keep the old hand operated line to fall back on when the main plant is out of action. We have also reduced our flexibility, because previously if we wanted to change the time of treatment in a bath we had only to instruct the operator. Now even a minor change in the timing means a modification to the mechanism.

We could re-establish the flexibility by installing a small computer to control the insertion and withdrawal of batches. By suitably programming, we could call for practically any time sequence we might need. Thus output and flexibility would be excellent, but in addition to failures of the mechanism we now have occasional computer failures and our overall reliability has dropped appreciably. We must certainly keep that old hand line in the corner, in order to process urgent orders when the main plant is out of commission!

The following examples illustrate the way in which the complexity of modern equipment, perhaps more than anything else, has led to the reliability problems which we have today.

System	*Approximate number of parts*
Domestic television	500 upwards, depending upon whether black and white or coloured
Large data handling equipment	250,000 upwards
Manned space mission	Several million parts

Thus the problem which manufacturers constantly have to face is not so much that the parts they use are unreliable, but that there are so many of them that there is always an appreciable risk that one somewhere is giving trouble.

In the above we have used the term parts, but the argument is quite general and applies equally to elements, units, blocks, sub-assemblies, etc.

6.3. parts in series

Suppose we have a system consisting of a number of parts and:

1 We know the reliability of each part.

2 Every part is vital in the sense that, if one fails, the whole system will fail.

To illustrate this problem, consider the transformer and rectifier sets, used to convert mains electricity to a suitable voltage and wave form. Suppose for simplicity that each part in each set has a reliability of 0·9, as defined in section 1.2. (In practice, the reliability of each part would be higher than this.) If we require only a transformer and nothing else, then clearly our system will have the same reliability as the one part it contains. Therefore:

$$\text{For 1 part} \quad \text{Reliability} = 0{\cdot}9$$

If however we also require a rectifier, then we have two things which can go wrong. Therefore:

$$\text{For 2 parts} \quad \text{Reliability} = 0{\cdot}9 \times 0{\cdot}9 = 0{\cdot}81$$

The reader may be surprised that the overall reliability is not 0·80. Since each part has a 0·10 or 10 per cent chance of failing, we might expect two parts to have 2×10 per cent equal 20 per cent chance of failing, and therefore an 80 per cent chance of working. The explanation is that there is 1 chance in 100 (i.e.

$0{\cdot}1 \times 0{\cdot}1$), that both parts will fail in the same system. The chances of failing thus consist of:

10 chances in 100 that the first part will fail, and in 1 of these chances the second part will also fail, plus the remaining
9 chances in 100 that the second part only will fail, making
19 chances in 100 of failure, and therefore
81 chances in 100 of successful working of the complete system.

Suppose now that we add further parts to our set, each of which is 0·9 reliable. We then get:

Number of parts in system	*Reliability of system*
1 part	$= 0{\cdot}90$
2 parts	$(0{\cdot}90)^2 = 0{\cdot}81$
3 parts	$(0{\cdot}90)^3 = 0{\cdot}73$
4 parts	$(0{\cdot}90)^4 = 0{\cdot}66$
5 parts	$(0{\cdot}90)^5 = 0{\cdot}59$
10 parts	$(0{\cdot}90)^{10} = 0{\cdot}35$
20 parts	$(0{\cdot}90)^{20} = 0{\cdot}12$
100 parts	$(0{\cdot}90)^{100} = 0{\cdot}000\ 026$

Now a reliability of 0·000 026 is merely a polite mathematical way of saying that there is really no hope at all that the system will work satisfactorily for the required time.

However, a system with only 100 parts is very simple by modern standards, but the part reliability we assumed is also very low. Suppose therefore we increase this latter to 0·999 or 99·9 per cent reliability. We then get:

Number of parts in system	*Reliability of system*
1 part	$= 0{\cdot}999$
2 parts	$(0{\cdot}999)^2 = 0{\cdot}998$
50 parts	$(0{\cdot}999)^{50} = 0{\cdot}952$
100 parts	$(0{\cdot}999)^{100} = 0{\cdot}906$
200 parts	$(0{\cdot}999)^{200} = 0{\cdot}820$
400 parts	$(0{\cdot}999)^{400} = 0{\cdot}673$
700 parts	$(0{\cdot}999)^{700} = 0{\cdot}500$

Thus with 700 parts, each 99·9 per cent reliable, there is only a 50 per cent chance that the complete system will be reliable. However, complex systems may contain thousands or even millions of parts, but fortunately our example is somewhat artificial since :

1 Very reliable parts are now produced, as the table in section 6.7 shows.
2 It is unlikely to be essential for every part to function correctly in order to ensure that the system as a whole works acceptably. A car for example, could hardly be considered to have failed because the bulb in the dash board light would not work.
3 Critical parts can be duplicated, so that if a failure occurs there is a spare to take over. This is called redundancy, and we shall consider it in detail in Chapter 8.

Where a number of parts are all essential for the satisfactory operation of a system, they are said to be in *series*. The terminology is analogous to that for a number of electrical components connected in series, as shown in the top section of Fig. 6.3(a). If the circuit is broken by any component, no current will

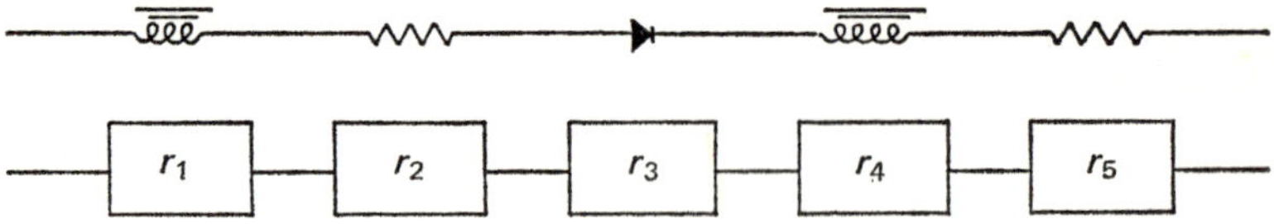

FIG 6.3(a). Parts in series

flow. In the same way parts in series in the reliability sense are drawn as shown in the lower half of Fig. 6.3(a), even though their physical relation to each other may be quite different. This merely indicates that if one part fails, the whole system or section thereof fails.

We can now express the relationship for parts in series in general terms.

Let the reliability of each part be $r_1, r_2, r_3, \ldots, r_n$
Then the reliability of the whole is given by:

$$R_s = r_1 \times r_2 \times r_3 \times \ldots \times r_n \tag{6.3a}$$

6.4. assistance from the bath-tub curve

However much we strive to use as few parts as possible we are going to be left with some, and in many cases a very considerable number of them. Therefore it is essential to choose parts which have the highest possible reliability, bearing in mind cost and other considerations. The bath-tub curve in Fig 2.1(a) summarizes our problem. Each part will have some pattern of failure rate against time, not necessarily bath-tub in shape, although this is a useful concept from which to start. Broadly our problem is to avoid early failures OA, replace before wear out failures BC and then to minimize failures during the constant failure rate period AB. In light-hearted terms, we must cut off both ends of the bath tub and lower its bottom. Since the methods by which part and component reliability can be improved are not confined to a particular section of the bath-tub curve, we will discuss them in general in the following sections.

6.5. choose parts which are inherently reliable

It often happens that a particular job can be done by more than one type of part. If for example we require a rectifier, we may have a choice between:

1 A vacuum diode and,
2 A semiconductor rectifier, for example copper oxide, selenium, germanium, silicon.

Reliability will probably prompt us to reject the vacuum diode, which will break if dropped and which requires a heater current, since this is something else to go wrong. We must therefore choose between the four semiconductor rectifiers. Of these, the now largely superseded copper oxide rectifier is surprisingly reliable, especially when it comes to reverse overloads. It has fallen from favour because for a given physical size it is only able to handle a small amount of power compared with the others. On this score silicon usually comes out best, followed by germanium.

Thus we must select our parts, bearing in mind the reliability and performance of each of the alternatives.

6.6. choose standard rather than specially designed parts

When a part or equipment is first designed and built, there are nearly always teething troubles. The designer and manufacturer cannot hope to foresee all the potential snags, and some of these may be discovered as a result of failure in service. Action will then be taken to put things rights.

When therefore we have to choose between a standard and a specially designed part, we are in effect choosing between a part whose snags have been largely discovered and corrected and one whose snags are likely to occur in our system. In practice the choice is not as easy as this. It may be that the performance to be expected from a specially designed part is much better than a standard part could achieve. In this case we may decide to design a part which is not unnecessarily non-standard, and then to subject early production to adequate tests.

It is useful to have a coding system, such as the Brisch or Opitz, by which a designer can easily locate all existing designs within the company which are similar to the one he needs, because he can then see whether any will do. Without such a coding system, it may be very difficult for a designer to ascertain what already exists, and then it may actually be easier to create a new part rather than spend time trying to locate an appropriate existing part.

6.7. use available performance and reliability data

Manufacturers of parts frequently offer their customers data sheets showing failure rates, and graphs of deterioration of performance with age, etc. and full use should be made of these.

Some care is necessary, however, because the environmental and other conditions under which we intend to use the parts may be very different from those envisaged by the manufacturer and simulated in his tests. We may have to supplement his data with actual tests, although as we show in section 18.1 rig tests are expensive and we should not embark upon them without careful consideration.

The following table gives some typical failure rates, but they are only intended as a general guide, since designs are continually changing and it is better to get up to date data which is relevant to the actual part concerned. Notice that failure rates are quoted, rather than reliability. As we said in section 2.7, failure rates for many electronic parts are approximately constant once the early failure period has passed, and it is then more convenient to use failure rates because they are not related to any particular life span. The reliability for any given operating period $= t$ can then be calculated from equation (3.2b), which was:

$$R(t) = \exp(-\lambda t)$$

Because the failure rates of parts are usually low, they are normally quoted in percentage failure per 1000 hours. Hence the figures given below must be divided by 100,000 to make them exactly equivalent to the definition of failure rate given in section 1.6.

Part	*Typical failure rate in % per 1000 hours*	
Capacitors		
Ceramic	0·01–0·05	
Electrolytic	0·04–0·3	depending on type
Glass and mica	0·01–0·04	
Paper	0·01–0·05	
Connections		
Soldered, wrapped, etc.	0·000 1–0001	
Inductors	0·01–0·08	

Integrated circuits		
Worth investigating; they can be very reliable.		
Rectifiers		
Germanium	0·08	
Silicon	0·005–0·05	
Relays	0·01–0·1	upwards
Resistors		
Composition	0·005–0·2	
Film	0·002–0·1	depending on type
Wire wound	0·01–0·2	
Transformers	0·01–0·04	
Transistors		
Germanium	0·01–0·1	
Silicon	0·005–0·1	
Valves	1·0–5·0	

6.8. buy from reliable suppliers

Parts which we do not make ourselves must be bought from a reliable supplier. Past experience is a great help, and an efficient goods inwards inspection system will be a valuable source of information. If we operate some sort of vendor rating, it will be relatively easy to compile a list of preferred suppliers, showing the particular types of part for which each has proved reliable. The latter is necessary because a supplier who is very good at making one part may be quite mediocre at making something else for which his experience is limited. We shall discuss this again in section 12.7.

6.9. derating

When we buy functional items, the manufacturer usually quotes a recommended rating, for example:

Safe working load = 10 tons,
3 watt resistor.

There is however nothing inherent about these ratings, as we can show by considering the problem of rating a mechanical hoist. No doubt the manufacturer will test prototypes until failure occurs, and this will give him an upper limit. He must then reduce this to what the hoist will safely bear, allowing for various additional factors, for instance:

1 It is practically impossible to completely eliminate internal flaws in the load bearing parts, so allowance must be made for the extent to which they may be weakened.
2 Misuse, including occasional overloading by the customer, will inevitably occur from time to time.

These considerations must be integrated into some sort of safety factor, from which the manufacturer will quote a safe working load. The lower he sets this rating, the safer he is, but if he sets it too low then the hoist will appear to the customer to be big, clumsy and expensive for the small load it will lift and it will be compared unfavourably with similar models from competitors. Hence a rating must be chosen which is as high as possible consistent with safety.

With electronic parts, the best rating is even more difficult to determine. When the applied voltage is very low, some parts do not work satisfactorily. Above this there is usually a useful working range within which the rate of deterioration is negligible, provided that the parts have been well manufactured. As the load is increased, however, the rate of deterioration also increases, usually at an ever greater rate. For many parts the process is complex, and deterioration appears to be due to a combination of:

1 The voltage and current themselves.
2 The heat they generate.

Thus the rate of deterioration can be reduced if we arrange to remove the heat which is generated quickly enough to restrict the resultant temperature rise. Improvement is limited, however, because the voltage and current cause deterioration in their own right. Alternatively, if we are satisfied with the reliability of the parts, we might use better heat dissipation to increase the electrical rating. Limitation would then come when the parts would no longer stand either the applied voltage or the current passing through them.

The problem is further complicated by factors such as time and wave form. Many devices will accept reasonably large overloads, provided they only last for a very short time. Voltages with a very peaky wave form, however, may cause breakdown when the rms value is quite acceptable.

Again the rating set by the manufacturer is likely to be a compromise. If he gives his parts too high a rating, the failure rate will be correspondingly high and he will receive an excessive number of customer complaints. If he sets it too low, his product will appear inferior when compared with that of his competitors. Probably he will set his rating as high as he dares.

Clearly then the designer has some discretion about the rating of a part he uses. If he wants to improve its reliability, he can assign a rating which is below that quoted by the manufacturer. Thus, the parts for a recent British satellite were derated so that in the worst conditions they received no more than 30 per cent of their maximum power rating and 50 per cent of their peak voltage rating. The designer must, of course, balance the improvement in reliability against the increased cost, weight and volume incurred by derating.

6.10. safety margins

Suppose we have a supply of a particular part which we test to breaking point under otherwise normal working conditions. This might, for example be a tensile test on a mechanical link, or a voltage overload on an electronic component.

If all the parts were identical and if the conditions of test were absolutely constant, then every part would have exactly the same ultimate strength, say M_B in Fig. 6.10(a). In practice of course this is never the case. There are always manufacturing differences

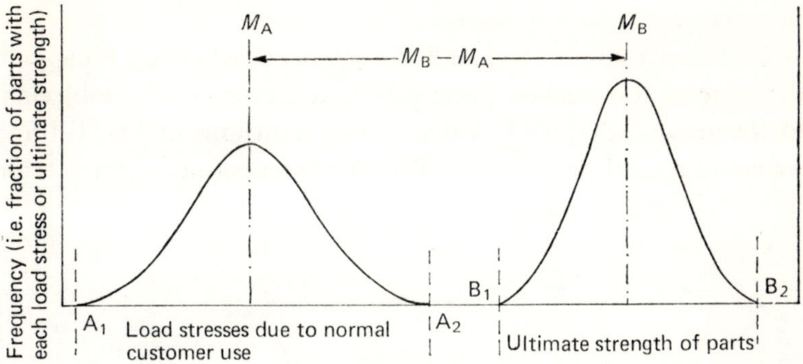

FIG 6.10(a). Safety margins

from one part to another, and there will also be variations in the conditions of test, even though these may be relatively small. Hence some parts will be stronger than M_B and some weaker. When we have enough information, we can plot a curve showing the fraction of parts with each ultimate strength, and this might look like curve B_1 to B_2 in Fig. 6.10(a). If the manufacturing process enjoys good quality control, then it is likely that all the variables in it will be small, and just as likely to be positive as negative. In that case, we shall have a situation which is mathematically analogous to that discussed in section 3.5 and a Normal frequency distribution curve will result.

Further to the left in Fig. 6.10(a) there is a corresponding frequency distribution curve, representing the variations in load stress during customer use. This is a much more difficult curve to predict. We may think we know the treatment our customers give, but there are always occasions when it is worse than expected. It is reasonable to suppose that variations in customer treatment will be greater than manufacturing variations, and therefore the spread from A_1 to A_2 is shown as greater than that from B_1 to B_2. It may or may not be Normal in shape.

To avoid any failures in service, we must ensure that the weakest part made is a little stronger than the worst the customer is likely to do to it, which means that B_1 must be just to the right of A_2. In Fig. 6.10(a) this is the case, but we might find the situation shown in Fig. 6.10(b).

In the area shown black, failures will occur whenever it happens that one of the weakest parts goes to a customer who subjects it to the most arduous of duties. Thus, returning to Fig. 6.10(a), we could regard the quantity $(B_1 - A_2)$ as a measure of the margin

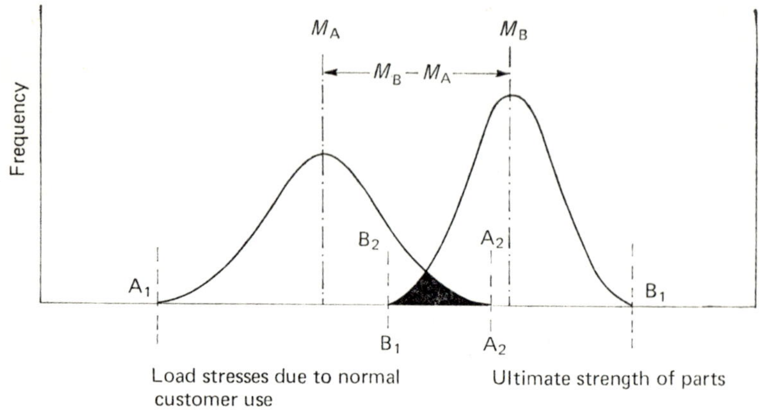

FIG 6.10(b). Safety margins, a situation in which some failures will occur

of safety. In practice, however, the points B_1 and A_2 are not particularly well defined, and so we usually use the mean values M_A and M_B. Some engineers then define the safety margin as:

$$\text{Safety margin} = \left[\frac{M_B}{M_A}\right] \quad \text{or} \quad \left[\frac{M_B}{M_A} - 1\right] \qquad (6.10a)$$

This however does not take account of the spread of the two distribution curves about M_A and M_B, and therefore it is usually better to proceed as follows.

1 Estimate the standard deviation for each curve, using equation (3.6c). Suppose these are:

σ_A for the curve of customer use.
σ_B for the ultimate strength of parts.

2 From the discussion in section 3.6, we know that in Fig. 6.10(a)

$$(A_2 - M_A) = 3\sigma_A \text{ approximately}$$
$$(M_B - B_1) = 3\sigma_B \text{ approximately}$$

Therefore $(M_B - M_A = k_A\sigma_A + k_B\sigma_B$ where k_A and k_B are constants. (6.10b)

Since we know the values of M_A, M_B, σ_A and σ_B, we can make this equation balance by setting suitable values for k_A and k_B. If we decide to make :

$$k_A = k_B = k, \text{ say}$$

then $$(M_B - M_A) = k(\sigma_A + \sigma_B) \qquad (6.10c)$$

and

$$k = \frac{(M_B - M_A)}{(\sigma_A + \sigma_B)} \qquad (6.10d)$$

Now we know that if k is less than 3, some failures are certain to occur, and in general the larger k the safer. Thus for extreme cases, such as manned space probes, values of k in excess of 10 may be used, but usually k is lower than this.

From equation (6.10d) we can see that in order to increase k, we must either increase M_B or decrease M_A, σ_A or σ_B. Consider what this involves.

1 INCREASE THE MEAN STRENGTH OF PARTS = M_B

To do this the average strength of parts produced in the factory must rise. It is the sort of improvement companies are constantly making, using new designs, new materials and new production methods.

2 DECREASE THE MEAN LOAD STRESS IN SERVICE = M_A

This is precisely what happens when a designer decides to derate parts, and it is therefore fully described in section 6.9.

3 DECREASE THE MANUFACTURING VARIATIONS EXPRESSED BY σ_B

In many cases, this is the best prospect for a quick improvement, yet it is often overlooked. In effect we are saying to the production staff : 'Your average quality M_B is satisfactory, but it is made up of a mixture. The best parts are excessively good, and will probably outlast the equipment into which they go, but the worst ones fail much too soon, and give rise to customer complaints. Can't you do something to make all your production much nearer to the average value M_B.' This is a job for the quality engineer, and we shall discuss how to tackle it in Chapter 12. Whether anything can be done depends largely on the inherent capabilities of the machines and processes involved.

4 DECREASE THE VARIATIONS IN CUSTOMER USAGE $= \sigma_A$

This is another area which often has possibilities over a period of time. In effect we want to avoid the worst treatment, bordering on misuse, which our customers sometimes impose upon our products. Possible ways of doing this are :

a Clear instruction manuals, written in the native language of the user, and illustrated as necessary. This is discussed in section 11.7.

b Customer training in the operation and maintenance of the equipment concerned.

c Good after-sales service.

6.11. burn-in

As the bath tub curve in Fig. 2.1(a) shows, parts are particularly liable to fail during the early part of their lives, thus giving rise to the early failure period. If the application is critical, we may consider it worthwhile to carry out tests in order to identify and remove these early failures. This operation is usually called *burn-in*. It consists of loading all the production parts concerned to their full rating, under normal working conditions, and continuing this treatment until all the early or burn-in failures have been removed. This often takes a few hundred hours; for example 250 to 300 hours is quite typical.

Inevitably 100 per cent burn-in tests are expensive; 250 hours represents about $10\frac{1}{2}$ days, assuming that the test is continuous day and night. If the quantities are at all large, we may have great difficulty in providing sufficient space on the test rigs.

6.12. selection of the best parts

Even after we have done all the foregoing, we may occasionally still be in trouble in terms of performance, if not reliability. In this case, it is sometimes possible to test batches and select the best parts. Whether this is an economic proposition or not depends not only on the advantage to be gained from it, but also on whether a useful application is available for those parts which are rejected. In some cases an instantaneous test can be used to detect parts with potentially bad ageing properties. Thus for a given method of manufacture, selenium rectifiers with a low initial forward resistance tend to age (i.e. change with time and use) less than those with a higher initial forward resistance.

CHAPTER 7 DESIGN FOR SYSTEM RELIABILITY

7.1. principles

In Chapter 6 we agreed that in order to achieve maximum reliability, we must have reliable parts and use as few as possible. In Chapters 7 to 11, we now consider how to unify them into a reliable system. In this chapter we shall deal with general principles, and then go on to study :

Chapter 8 Redundancy.
Chapter 9 The environment in which the system must work.
Chapter 10 How to make it easy to operate.
Chapter 11 How to make it easy to maintain.

Our starting point is the customer or market specification which we considered in Chapter 5. This should tell us the required reliability, possibly in units of failure rate or MTBF, and in addition give full details of performance, safety, environment, cost, etc. In this book we are looking specifically at reliablity, but the achievment of reliability at the expense of everything else is seldom difficult. It is getting every requirement into one system which presents difficulties. Equally we must not assume that because high priced systems, such as aircraft, must be very reliable, low priced systems can have a correspondingly low reliability. Sometimes this is true, but the steering mechanism of a car, for example, must be both relatively cheap and very reliable.

7.2. blocks or units in series

Every designer will have his own way of starting a design, depending partly on what he is designing. The method however will probably follow roughly that discussed in section 5.4. It is usually convenient to start with a block or unit diagram, in which we break the whole system down into reasonably self-contained sections and to each of which we can assign performance and reliability requirements.

Suppose we have to design a piece of automatic equipment, to carry out an electrical test on the output of an industrial process. Our block or unit diagram might be as shown in Fig. 7.2(a).

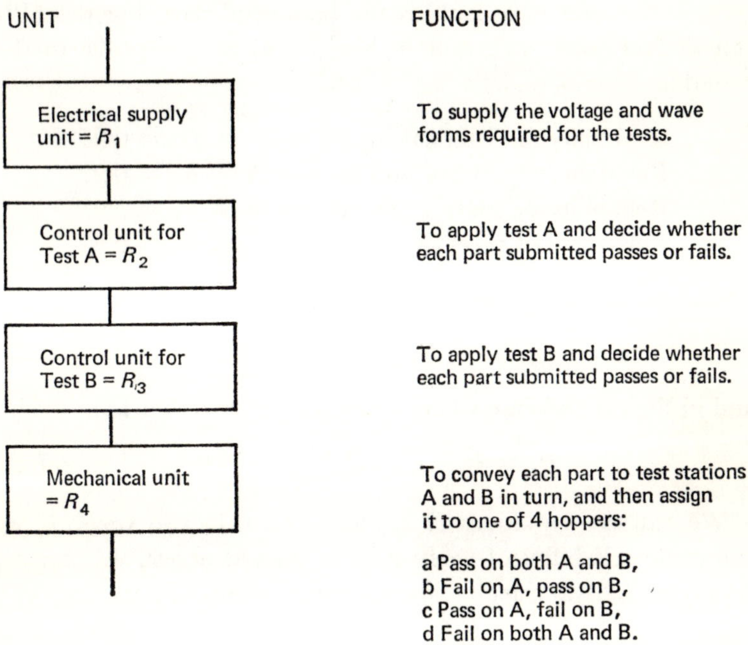

FIG 7.2(a). Block or unit diagram for an automatic test equipment

There are four units, which in terms of their function are separate and identifiable. Of course, if we see the equipment in use it may not look like this at all. Probably the most prominent feature will be the machine itself (i.e. the mechanical unit) and somewhere alongside will be a cabinet containing all the electrical equipment of the other 3 units.

If we know the reliability required for the overall system, we can proceed to allocate reliability targets to each of the 4 units. The method for doing this is a logical development of equation (6.3a). We said that from a reliability point of view, parts could be considered to be in series if the failure of one part would cause the whole to fail and that the reliability of the system would then be given by:

$$R_s = r_1 \times r_2 \times r_3 \times \ldots \times r_n \qquad (6.3a)$$

However we made no assumptions about what we meant by a part, so that our part could really be several parts together. We can in fact treat each unit in Fig. 7.2(a) as susceptible to this equation. Therefore, let:

Reliability of electrical supply unit $= R_1 = 0{\cdot}95$
Reliability of control unit for test A $= R_2 = 0{\cdot}92$
Reliability of control unit for test B $= R_3 = 0{\cdot}93$
Reliability of mechanical unit $= R_4 = 0{\cdot}80$

Then the reliability of the whole system is:

$$R_s = R_1 \times R_2 \times R_3 \times R_4 \qquad (7.2a)$$

and in the general case where there are j units or blocks in series:

$$R_s = R_1 \times R_2 \times R_3 \times \ldots \times R_j \qquad (7.2b)$$

We can use equations (7.2a) and (7.2b) in two ways. If we know the reliability of each unit, as quoted above, we can use equation (7.2a) to calculate the system reliability. Thus:

$$\begin{aligned} R_s &= 0{\cdot}95 \times 0{\cdot}92 \times 0{\cdot}93 \times 0{\cdot}80 \\ &= 0{\cdot}65 \end{aligned}$$

Alternatively, if the customer or market specification tells us what the complete system reliability must be then we can allocate minimum reliabilities to each unit. We shall not necessarily give each unit the same share of the overall reliability. Even without precise figures, we may guess that the electrical supply unit will be very reliable, unless perhaps it must deliver a complicated wave form. On the other hand, the mechanical unit is sure to break down from time to time, and so we shall allocate it a lower minimum.

It may be that from past experience, we know roughly what the reliability of each unit is likely to be, and we also know what reliability the customer or market specification demands for the overall system. If this demand turns out to be less than $R_1 \times R_2 \times R_3 \times R_4$, we have a bit in hand, but if it is more, we are in difficulties. We can either persuade the customer to reduce his demand or we must find ways of increasing the reliability of the units until their product at least equals the required value of R_s.

Notice that R_s at 0·65 is lower than the reliability of any of the units which make up the system (i.e. 0·95, 0·92, 0·93 and 0·80). This is as we should expect, because the overall reliability must take account of the failures in all four units put together.

7.3. the black box

It is convenient for many purposes to consider each unit or block as a *black box*. This is a term which is popularly used to describe the flight recorder on an aircraft, but its real meaning is much wider. By a black box, we mean part of a system which can be considered as a separate and identifiable unit. We know what its input is, what it does and hence what its output is, but we do not worry about the parts it contains or how they work. It is as though the whole were sealed in a black box whose contents are a complete mystery. Thus, for some purposes, the storage unit of a computer can be considered as a black box. We know what data is fed into it and stored and that when required that same data can be retrieved and fed out again, but we do not worry about what is inside the storage unit to enable it to do this.

7.4. the addition of failure rates

In sections 2.7 and 2.8, we noted that failure rates for many electronic parts and systems tend to be constant and, where this is the case, they can be added together to estimate the failure rate of the unit, etc. concerned. For example, if a system consists of two units in series, where one breaks down twice a year and the other three times, then clearly the whole system fails five times a year. The calculation is further simplified, because a unit usually contains a number of parts of the same general type. Thus suppose that the electrical supply unit of section 7.2 has a total of 29 parts, as shown below in columns A and B.

A *Type of part*	B *Number in unit*	C *Failure rate per part (in % per 1000 hours)*	D *Total failure rate for that type of part (in % per 1000 hours)*
Transformer	3	0·02	0·06
Silicon diode	6	0·01	0·06
Variable resistors	2	0·20	0·40
Switches	6	0·02	0·12
Inductors	6	0·05	0·30
Capacitors	6	0·01	0·06
Totals	29		1·00

Opposite each in column C we write the failure rate per part, expressed as percentage failures per 1000 hours. Next we multiply each failure rate by the number of parts of that type to give column D, which we total to get the overall failure rate of the unit.

This shows that the electrical supply unit will have an overall failure rate of about 1·00 per cent per 1000 hours, which is an MTBF of 100,000 hours. If we want to know the reliability over say 5000 hours, assuming that the failure rate remains constant, we use equation (3.2b):

$$\begin{aligned} R_1 &= \exp(-\lambda t) \\ &= \exp\left(-\frac{1\cdot 0}{100{,}000}\times 5000\right) \\ &= \exp(-0\cdot 05) \\ &= 0\cdot 95 \text{ as already quoted in section 7.2} \end{aligned}$$

In calculating reliabilities, we must allow for the fact that some units may not be operating all the time. A unit, for example, may only be used when the main unit is started up. Thus the self-starter of a car is only used to start the engine, and t in equation (3.2b) must be reduced accordingly. Sometimes it is convenient to use 'number of cycles' or 'number of operations' as the unit of time. The latter might apply to the operations of a relay.

7.5. areas of knowledge and ignorance in reliability design

In considering the design of our automatic test equipment in sections 7.2 and 7.3, we assumed that all the information we needed was readily available, but this may not always be true. There are nearly always areas where past knowledge and experience is available, so that we can predict with confidence what the reliability of the unit concerned is likely to be. By elimination, we can then identify the areas of ignorance, where little is known, and these will need special attention. We may wish to design new parts or to utilize new materials which have just become available in order to achieve the required performance. Unfortunately, information about their reliabilities may be lacking, and we may have to carry out tests in an effort to fill the gap. These must be planned with care, because tests are nearly always expensive, and if not properly devised and executed may be misleading as well. Adequate information must be obtained however. At the design stage, the manufacturing drawings and specifications can be altered relatively easily, but once manufacture is under way, that same alteration may be very expensive. It is therefore vital to ensure that the design is a satisfactory practical proposition before it is released for manufacture. It is not enough for a prototype to 'get by' when built with great care by a highly skilled team. We must consider whether it will still be all right when built by factory personnel and equipment.

Clearly, the more relevant information which is available the better, and for this reason some firms have established a bank of reliability data for their own use. This might include :

1 Preferred circuits, for which the performance and reliability characteristics are known.

2 Preferred parts, components and other equipment.
3 Preferred suppliers and sub-contractors.
4 A handbook of standard procedure and similar information.

Sometimes there appear to be several alternative design possibilities, each of which has the potential to satisfy the customer or market specification. If we have full information about each, we may be able to do a fairly direct comparison. Thus we can draw a block or unit diagram for each and compare their reliabilities, performance, cost, etc. Often, however, available information is incomplete, and we may have to do an initial investigation and appraisal on each of them. The decision may be a difficult one. Design A perhaps is reasonably certain to achieve a known reliability, but unfortunately this reliability barely satisfies the customer specification. Design B on the other hand shows promise of achieving a higher reliability, but being a new design there is the ever present risk that difficulties may be experienced with it. If the risks can be expressed as mathematical probabilities, it is possible to calculate which design is to be preferred. The problem is however that the decision so obtained depends not only upon the calculation, which is probably correct, but also upon the values selected for the various risks, which may be very inaccurate. Inevitably some considerable degree of judgement is necessary, and this may also prove to be inaccurate.

7.6. interfaces

As we mentioned in section 2.8, the assembly of parts brings additional risks of failure at the interfaces between them. Soldered joints in particular are by no means infallible, and failure rate curves for parts are invariably for the parts themselves, and not for the efficiency or otherwise of the way they have been soldered into the circuit. A further snag is that the heat from the soldering iron, if not correctly handled, may damage the part itself. One of the many advantages of the use of techniques such as printed circuits is that failure at the interfaces becomes less likely.

Multi-pin plug and socket connections enable a unit or element to be replaced very quickly, but unfortunately they may be a further source of unreliability. Too often an equipment failure proves to be due to a faulty connector, or to one which was not pushed fully home.

In principle the above means that the interface should be treated as though it was a part with its own reliability. Thus if two parts in series have reliabilities of r_1 and r_2, and if the interface has a reliability of r_{12}, then the overall reliability of that element is:

$$R = (r_1) \times (r_{12}) \times (r_2) \qquad (7.6a)$$

7.7. dealing with variations in parts

Our equipment when built must not only work now but must continue to work throughout its planned working life. Even when new however, it is unlikely to be exactly at its nominal value. Thus a nominal 100 Ω resistor is unlikely to be precisely 100 Ω, because of variations in its manufacture. If we are prepared to pay for them, we can buy resistors with a very small manufacturing tolerance, but even these will vary a little. The mechanical engineer is no better placed. All parts vary slightly in size, and imposed on this will be variations in quality such as hard spots, porosity, slag and sand inclusions and differences in chemical composition.

There is no one solution to these variations, although if a system consists of many parts it is almost certain that they will tend to cancel each other out. There may be a residual effect, however, with which we shall have to cope, possibly by providing means of adjustment such as voltage taps on transformers, variable resistors, mechanical adjustments, etc. Sometimes a designer will do a 'worst case' analysis, in which he calculates how his equipment would perform if all the tolerances were unfavourable. Only in a simple case is there usually any real risk that all the parts will be at the worst end of their tolerance band, in the same unit. Where there are large numbers of parts, it is almost impossible for this to happen. While some tolerances may be unfavourable, others will be favourably placed. In such cases it may be more realistic to try to identify the parts which would be critical if they were at the extreme of their tolerance band, and consider what would then happen to the performance of the equipment.

7.8. marginal testing

A useful technique for checking these extreme conditions is known as *marginal testing*, and a simple example of this is shown in Fig. 7.8(a). Suppose a particular circuit will only operate satisfactorily if its current is between 50 and 75 mA. A range of voltages is applied, and for each the current in the circuit is noted. A curve AB is then plotted as shown. Two vertical lines are drawn at 50 and 75 mA, and where each touches the curve two horizontal lines are added. These show that the permissible variation in the applied voltage is from 16 to 22 V. The test will have to be repeated with various critical parts at the extremes of their tolerance bands, in order to decide how much the voltage can vary. This is likely to be rather less than the 16 to 22 V of Fig. 7.8(a).

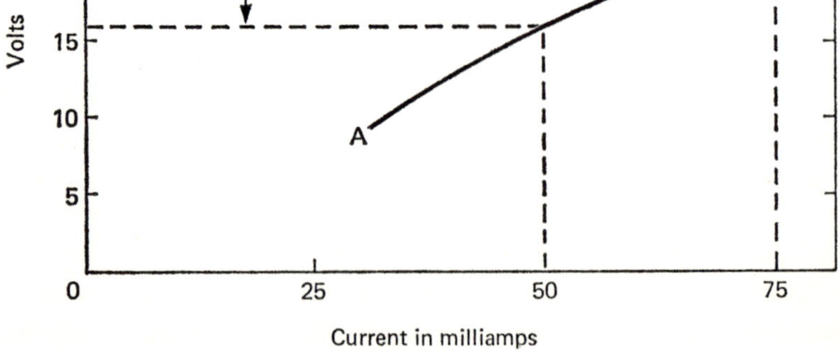

FIG 7.8(a). Example of marginal testing

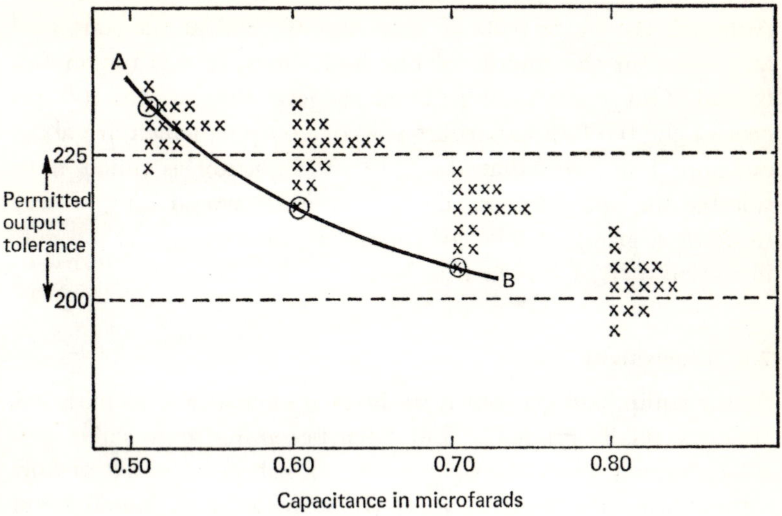

FIG 7.8(b). Selection of a suitable capacitance

The above is a very straightforward case. More typically, there would be many checks to do and it might be desirable to simulate the results on a computer.

We can use a modification of the above technique, to select suitably rated parts for a circuit. Thus suppose that a supply of capacitors is available with nominal values 0·50, 0·60 and 0·70 μF, respectively, and we wish to decide which to specify, so that a particular circuit has an output of between 200 and 225 cycles per second. Again we draw a graph, this time relating capacity in microfarads to the output in cycles per second and add two horizontal lines at 200 and 225 cycles per second. If we now merely test one capacitor of each rating, we should again be able to draw curve AB as shown and on the basis of this the 0·60 μF capacitor is best, because it gives an output in the centre of our two limit lines. However we have taken no account of the fact that there will be a variation in performance within each capacitor rating due to manufacturing variations in the capacitors themselves. We must therefore repeat the check, using a reasonable number of each. The points so obtained are then plotted as shown in Fig. 7.8(a), and we now come to a different conclusion. Although the single 0·60 μF capacitor we had at the start gave an output in the middle of our limit lines, it was not in fact typical. Our further information suggests that at least 60 per cent of the 0·60 μF capacitors will give outputs which are above our top limit, and that the 0·70 μF capacitor is much more suitable for our purpose. To be certain, it would now be wise to check a group of 0·80 μF capacitors, as shown, and this confirms that 0·70 μF is in fact the best.

7.9. insensitivity

As the equipment gets older, so the characteristics of its parts will change, usually gradually, but often becoming appreciable over a long period. Since the equipment will still be expected to work satisfactorily, the designer must try to make it as insensitive as possible to these changes. Where necessary, he may overcome the difficulty by providing means of adjustment.

He must also try to arrange that the failure of one part does not damage others. Thus if one part becomes a short circuit, this should not overload other parts in the system.

If, when the design is over-sensitive, equipments are despatched to customers containing too many parts which are very close to their tolerance limits, then early failures may occur, thus adding to the left hand side of the bath-tub curve, as discussed in section 2.2.2.

7.10. design for both manufacture and test

The designer always has a double objective :

1 He must design to satisfy the customer.
2 His designs must be within the manufacturing and test capabilities of the factory concerned. It is no good drawing something which either cannot be made, or if it is made, cannot be checked to make sure that it is correct.

This emphasizes the desirability of consulting both the production engineer and the quality engineer, at the design stage.

7.11. the 'fail safe' principle

Since nothing is absolutely reliable, we must plan to minimize the effects if and when a failure occurs: In particular we must ensure that there will be no danger of people getting hurt, and as far as possible no damage to property either. This is the 'fail safe' principle.

Thus we should like an aircraft to be completely reliable, but since this is impossible a maximum permitted failure rate has to be set. But even then, it must fail safe. It must not for example, crash. The railways have a similar problem, in that the failure of any piece of equipment must not cause an accident. Signalling equipment illustrates this point, and a typical arrangement is shown in Fig. 7.11(a).

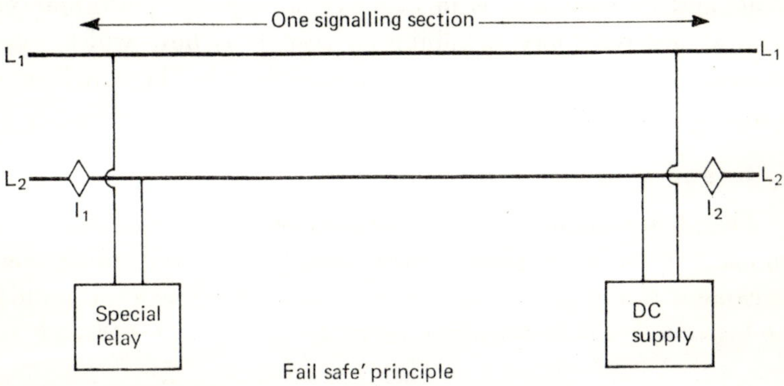

FIG 7.11(a). Fail safe applied to railway signals

The rails on which the train runs L_1L_1 and L_2L_2 are divided into convenient sections, and only one train is allowed into each section at a time. At the end of each section, the rail is electrically insulated from the next, by insulators I_1 and I_2. The signal can only show green when a special relay is energized, and this is done from a d.c. supply which must run along the whole length of the track in that section in order to reach it. Thus:

1 If the d.c. supply fails, or the circuit is somehow broken, the relay will not energize, and so the signal will show red.

This could delay a train, but should not cause an accident.

2 Once a train has entered a section, its wheels and metallic framework cause a short circuit between rails L_1L_1 and L_2L_2. The relay de-energizes, and the signal shows red. Hence another train cannot enter the section until the first has left it.

Further protection is provided against mistakes by the signalman, who cannot for example change the setting of signals and points ready for another train, while the first is still occupying the line concerned.

7.12. the use of value analysis

Value analysis originated in the USA during the Second World War, and its basic aim is to achieve the required performance, quality and reliability at minimum cost. It is now widely employed in this country, and it is characterized by the way it uses a small team as follows.

1 SELECT A DESIGN FOR STUDY

The choice might be a sub-assembly or even a component. It should not be a complete system, unless it is a very small one, because it would be impracticable to tackle it in one go. It would be better to break it down into units and tackle each separately.

2 DEFINE ITS FUNCTION

We must know exactly what our sub-assembly is intended to do, and as we said in section 5.2 this is usually better expressed as a service rather than as a description of the hardware. The latter may well get altered out of all recognition.

3 COLLECT INFORMATION

Every piece of information which could possibly be useful is collected, and this will include:

a The performance and reliability required.
b The cost of each part.

c Drawings and specifications.
d Materials used.
e Production methods.
f Quantities to be made.
g Markets which the product must satisfy.

4 SPECULATION

This is the most original part of the value analysis technique. A small team of say 3 or 4 people (for example the value engineer, the designer, the production engineer and the quality and reliability engineer) meet for a brainstorming session. If it is practicable, an example of the product to be considered is in the room, together with a set of all the parts of which it is composed. The team then start to make suggestions for reducing the cost, or improving the quality and reliability of the product concerned. One idea leads to another, and all are noted down. At this stage the 'rules of the game' forbid discussion of the suggestions made because this might stem their flow. Thus if someone says, 'If we made it in plastic, it would not corrode', someone else is not permitted to reply, 'Don't be silly, it's got to operate at 600°C'. Of course some of the ideas noted really will be impracticable, but among them will probably be some very sound suggestions.

5 EVALUATION

Next the ideas are carefully sifted and evaluated. Many will be rejected, but even so they may have been the very ones which led on to useful ideas later in the brainstorming session. Usually there are some ideas which are promising enough for further investigation.

6 VERIFICATION

These are then checked to make sure they are both practical and economical, while still achieving the required quality, performance and reliability. The best ideas are moulded together into a single proposal, which usually results in an improved design of the product.

7 PRESENTATION

Finally the proposals must be presented and 'sold' to higher management.

The foregoing is a very brief summary of value analysis, and of its sister subject value engineering. It would be inappropriate to enlarge here, but further information can be found in *Value Analysis and Value Engineering* by Frederick Oughten, published by Pitman.

CHAPTER 8 REDUNDANCY

8.1. types of redundancy

Even when we have done everything which we discussed in Chapters 6 and 7, we may find that we have still not achieved the overall reliability demanded. This may be primarily because some of the units which make up the system are insufficiently reliable. Hence we may decide to duplicate them, so that if one unit fails there is another similar unit there to carry on working, and so avoid failure of the whole system. This technique is called *redundancy* and, as a definition, it is the provision of more than one means of accomplishing a given function. The following are examples of the different ways it can be used.

1 On an aircraft, the pilot needs an accurate altimeter to tell him how high he is flying, and since this information is essential, provision must be made for the odd occasion when it goes wrong. The obvious possibility is to provide two identical meters, so that if one goes wrong readings can be taken from the other. This however presents a difficulty in that if each meter shows a different height, there is no way of telling which is correct. The aircraft therefore has three altimeters, so that if one goes wrong, the two which give the same reading can be identified as correct.

2 In the operating theatre of a hospital, if the mains electricity fails, provision is made to switch over to an emergency supply, often from a standby motor generator.

3 The spokes of a bicycle wheel illustrate a different form of redundancy, because usually several can break without a serious drop in performance or reliability. However a minimum number of spokes must remain, and they must be roughly evenly distributed around the hub. This is called partial redundancy.

4 Slightly different again, we have the spare wheel of a car. Modern tyres do not often puncture, but when they do a motorist without a spare wheel is stranded. Hence a spare is provided, so that the car operates satisfactorily so long as at least 4 of the 5 wheels are unpunctured.

There are two main types of redundancy:

1 ACTIVE REDUNDANCY Here all the alternative means of achieving a given function are energized whenever that section of the system is operating. Thus the altimeter is an example of active redundancy.
2 STANDBY REDUNDANCY Here one of the alternatives is energized at a time, and there is provision so that if one fails another can be switched in.

8.2. active redundancy

Suppose that one unit in our system is insufficiently reliable, so we decide to incorporate three units in active redundancy. They each perform the same function, and are designed so that the system will continue to operate so long as at least one of them is working. The three units may be identical, but as this is not necessarily so, we will designate the reliability of each by R_1, R_2 and R_3, as shown in Fig. 8.2(a). Notice that the units are drawn

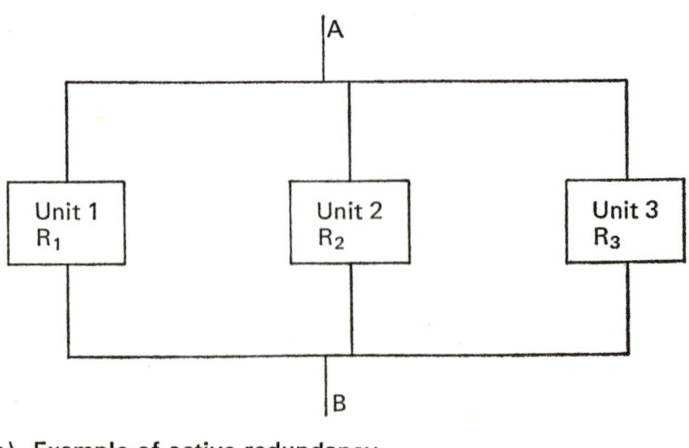

FIG 8.2(a). Example of active redundancy

as though they were in parallel in the electrical sense. The electrical analogy is that current will flow from A to B so long as there is an unbroken circuit through at least one unit. (It is convenient to use the term unit, but in fact the items in parallel can be anything from parts up to sub-systems.)

We shall need to be able to calculate the overall reliability of the units in parallel, and this is the probability that at least one of them is still working. This is cumbersome to do directly, and it is easier to calculate the probability that all three have failed, and then subtract this from 1·00 utilizing equation (3.2d). The steps are as follows.

1 We calculate the probability that each unit will fail:

Therefore:

$$\text{Probability that unit 1 fails} = F_1 = (1 - R_1)$$
$$\text{Probability that unit 2 fails} = F_2 = (1 - R_2)$$
$$\text{Probability that unit 3 fails} = F_3 = (1 - R_3)$$

2 Using a similar argument to that which led to equation (7.2b), we have:

$$\text{Probability all 3 units fail} = F_b = F_1 \times F_2 \times F_3 \qquad (8.2a)$$

3 From equation (3.2d), the probability the block of 3 units in parallel operates satisfactorily is:

$$R_b = (1 - F_b) = 1 - (F_1 \times F_2 \times F_3). \qquad (8.2d)$$

Therefore in general when there are k units in parallel

$$R_b = 1 - (F_1 \times F_2 \times \ldots \times F_k) \qquad (8.2c)$$

EXAMPLE

Suppose $R_1 = R_2 = R_3 = 0{\cdot}90$

Then probability unit R_1 fails $= F_1 = (1 - R_1)$

$$= (1 - 0{\cdot}90)$$
$$= 0{\cdot}10 = F_2 = F_3$$

Therefore:

Probability all 3 units fail $= F_b = F_1 \times F_2 \times F_3$

$$= 0{\cdot}10 \times 0{\cdot}10 \times 0{\cdot}10$$
$$= 0{\cdot}001$$

Therefore:
Probability block of 3 units operates as intended

$$= R_b = 1 - F_b$$
$$= 1 - 0{\cdot}001$$
$$= 0{\cdot}999$$

Thus, although one unit alone would only be 0·90 or 90 per cent reliable, three units in parallel are 0·999 or 99·9 per cent reliable.

8.3. improvement in reliability from active redundancy

Fig. 8.3(a) shows the improvement in reliability which can be obtained by putting up to 4 similar units in parallel. Notice that

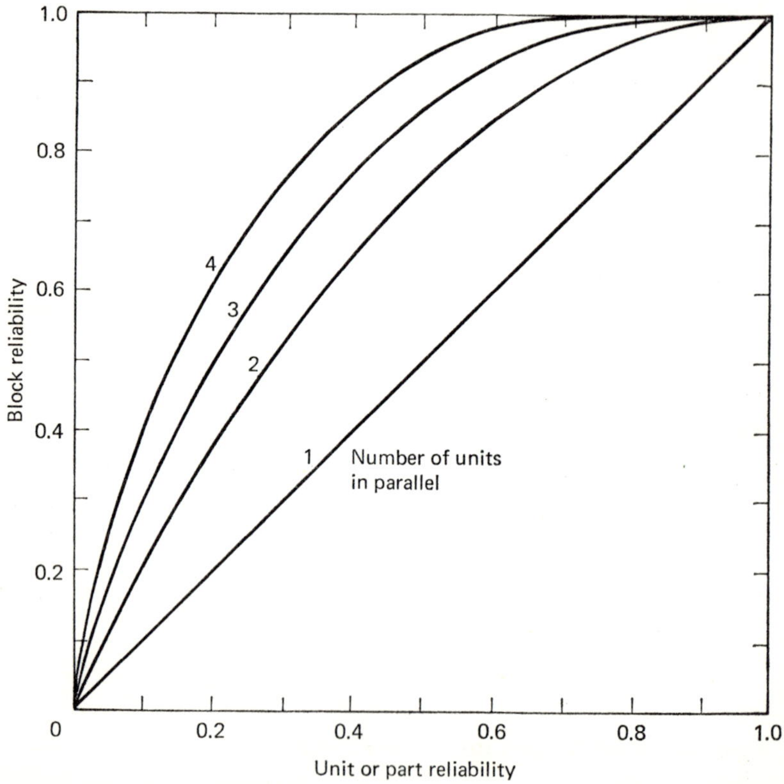

FIG 8.3(a). Improvement in reliability when units or parts are used in active redundancy

this improvement is not uniform, regardless of the reliability of the units. Thus:

1 Two units in parallel will in many cases give a useful improvement. If however we insert a third unit, the additional improvement is much smaller, and in general there is no advantage in putting large numbers of units in parallel.
2 We get the biggest increase when the unit reliability is around 0·35 to 0·50. If the unit reliability is already 0·85 or above, two units in parallel may be useful, but the use of three upwards gives little further improvement.
3 When the unit reliability is very low, it is difficult to get an acceptable block reliability from the use of redundancy alone. This is as we should expect, because if our units are unreliable the use of more of them only puts in more unreliability, and this offsets the gain from redundancy.

8.4. detection of a failed unit

One practical problem with active redundancy is that of detecting when a unit fails. Thus if two units are in active redundancy and one fails, the other will carry on working and the operator may not even know that a failure has occurred. This is dangerous, because the protection provided by redundancy has now gone. Some time later the second will also fail, and this time the system will fail with it. Therefore it is essential to have a detection device which will indicate when a unit fails, so that we can ensure that it is restored to working order.

An interesting example occurs in the lamps used for many railway signals. The lamp contains two filaments in parallel. One is of relatively low resistance, and this carries a fairly large current to light the signal lamp. The second filament is of higher resistance and, so long as the first is working, the current in the second is insufficient to light it appreciably. If however the first filament fails, then the regulation of the supply is such that the voltage across the second rises and so carries on the illumination of the lamp. It is however provided that even when fully lit, the second filament is not as bright as the first. Hence the maintenance man on his rounds will notice its dimness, realize that the bulb is now on its second filament and replace it.

8.5. the MTBF when active redundancy is used

If we have two units in active redundancy, the overall MTBF will not be doubled because both are energized together from the start, and so it depends upon how long before the second fails. If there are k identical units in parallel, each with a constant failure rate of λ, then the mean time between failures for the whole block is given by:

$$\begin{matrix}\text{MTBF of } k \text{ units in} \\ \text{active redundancy}\end{matrix} = \theta_b = \frac{1}{\lambda} + \frac{1}{2\lambda} + \frac{1}{3\lambda} + \ldots + \frac{1}{k\lambda} \quad (8.5a)$$

It is not necessary to know the derivation of the above equation, but for readers who are interested it is deduced as follows. Suppose we were able to put a large number of units on an infinitely long life test and observe the time when each failed. Then:

$$\text{MTBF} = \frac{\text{Total time all units survive on test}}{\text{Total number of units tested}} \quad (8.5b)$$

Since units fail from time to time, the fraction remaining on test will progressively decrease. Now the fraction which survive at any time t is given by the reliability $R(t)$ at that time. This value of reliability however will only be correct for an instant dt, before further units fail, and $R(t)$ takes on a new and lower value. Thus during the instant dt:

$$\begin{pmatrix}\text{Total time all units} \\ \text{survive on test} \\ \text{during instant d}t\end{pmatrix} = \begin{pmatrix}\text{Number surviving} \\ \text{at time } t\end{pmatrix} \times \begin{pmatrix}\text{Duration} \\ \text{d}t\end{pmatrix}$$

$$= \begin{pmatrix}\text{Number starting} \\ \text{the test}\end{pmatrix} \times \begin{pmatrix}\text{Fraction surviving} \\ \text{at time } t\end{pmatrix} \times \begin{pmatrix}\text{Duration} \\ \text{d}t\end{pmatrix}$$

$$= [k] \times [R(t)] \times [\text{d}t] \quad (8.5e)$$

In order to take account of all the time during which units are on test, we integrate from $t = 0$ to $t = \infty$. Therefore:

$$\text{Total time all units survive on test} = \int_0^\infty kR(t)\,\text{d}t \quad (8.5d)$$

Therefore from equation (8.5b):

$$\text{MTBF} = \frac{\text{Total time all units survive on test}}{\text{Total number of units tested}}$$

$$= \int_0^\infty \frac{kR(t)\,\mathrm{d}t}{k}$$

$$= \int_0^\infty R(t)\,\mathrm{d}t \qquad (8.5\text{e})$$

So far we have made no assumptions about how $R(t)$ varies, or how our units are arranged. Suppose therefore that we have two units in parallel, each having a reliability R and a constant failure rate of λ. Therefore:

$$\text{Reliability of block} = R_b = 1 - F^2$$
$$= 1 - (1-R)^2$$
$$= 2R - R^2$$

But from equation (3.2b),

$$R = \exp(-\lambda t)$$

and therefore

$$R_b = 2\exp(-\lambda t) - \exp(-2\lambda t)$$

Now insert this in equation (8.5e), where R_b replaces the general term $R(t)$. Thus:

$$\text{MTBF} = \int_0^\infty R_b\,\mathrm{d}t$$

$$= 2\int_0^\infty \exp(-\lambda t)\,\mathrm{d}t - \int_0^\infty \exp(-2\lambda t)\,\mathrm{d}t$$

$$= \frac{-2}{\lambda}\Big[\exp(-\lambda t)\Big]_0^\infty + \frac{1}{2\lambda}\Big[\exp(-2\lambda t)\Big]_0^\infty$$

$$= \frac{-2}{\lambda}[-1] + \frac{1}{2\lambda}[-1]$$

$$= +\frac{2}{\lambda} - \frac{1}{2\lambda}$$

$$= \frac{1}{\lambda} + \frac{1}{2\lambda}$$

and this corresponds to equation (8.5a), when there are only two parallel paths.

8.6. combination of series and parallel units

In practice a system may present itself as a combination of series and parallel units, and we may have to look carefully to determine its precise configuration. In effect, units will be in parallel whenever there is an alternative path which will enable the system to work satisfactorily, and they will be in series whenever it is essential for all the units concerned to work simultaneously.

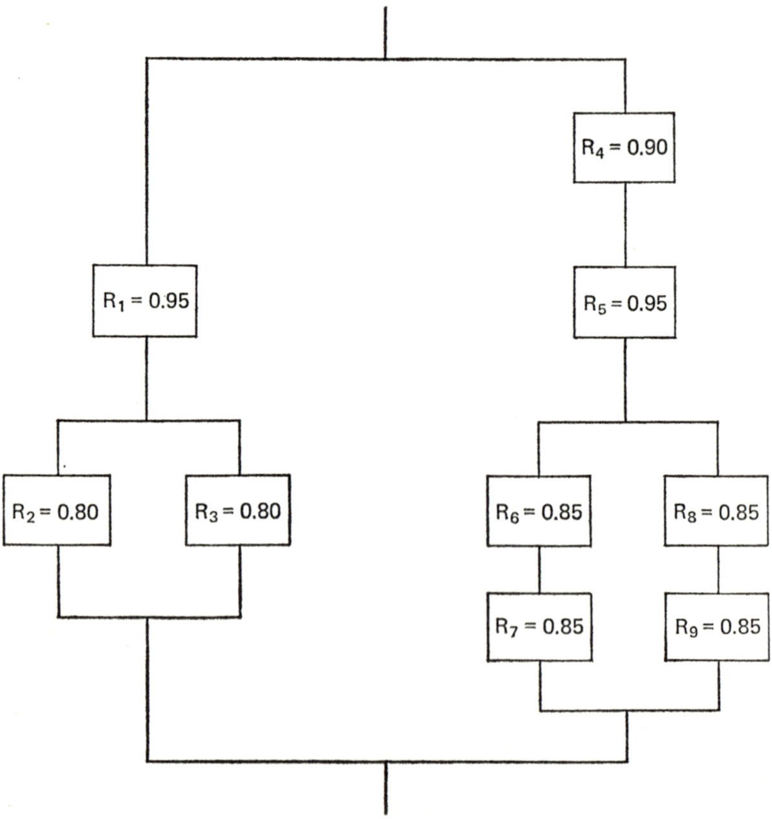

FIG 8.6(a). Example of a series/parallel arrangement of units

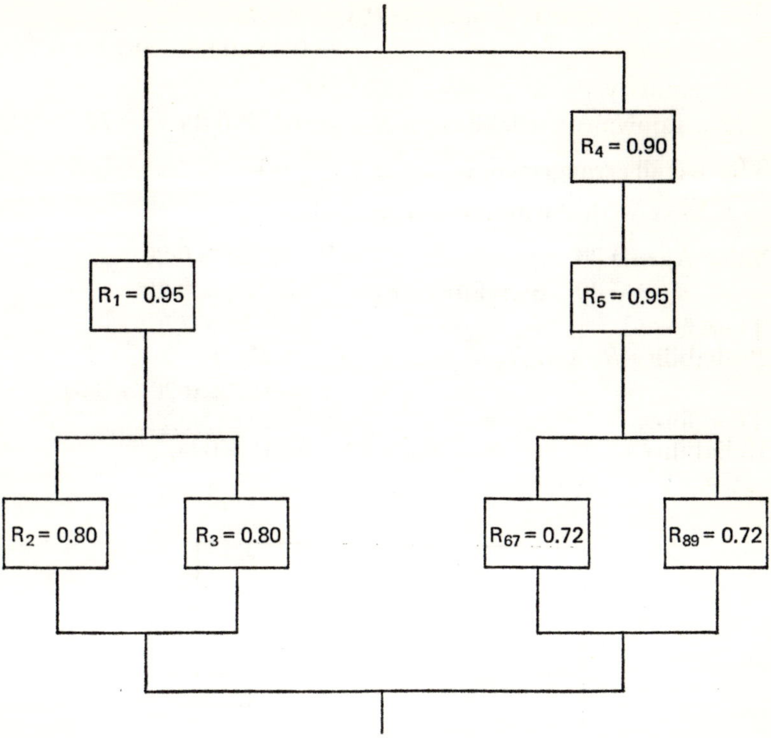

FIG 8.6(b).

Suppose we have a system as shown in Fig. 8.6(a). At first sight it looks formidable, but it is not difficult if we tackle it as follows:

1 Look for units in series within a parallel configuration, and calculate the reliability of a single equivalent unit, using equation (6.3a).
2 Reduce each parallel arrangement to a single equivalent unit using equation (8.2c).
3 Repeat (1) and (2) above, always dealing with the smallest recognizable configurations first.

Applying this to Fig. 8.6(a), we proceed as follows.

1 R_6 and R_7 are in series, within a parallel arrangement. From equation (6.3a) therefore:

Equivalent reliability $= R_{67} = R_6 \times R_7$
$= 0{\cdot}85 \times 0{\cdot}85 = 0{\cdot}72$

2 Similarly:
Equivalent reliability $= R_{89} = 0{\cdot}85 \times 0{\cdot}85 = 0{\cdot}72$

The overall arrangement is now as Fig. 8.6(b).

3 Next we deal with the two parallel sections.

Since $R_2 = 0{\cdot}80$ therefore: $F_2 = (1 - 0{\cdot}80) = 0{\cdot}20$
$R_3 = 0{\cdot}80$ therefore: $F_3 = (1 - 0{\cdot}80) = 0{\cdot}20$

Therefore:
Probability R_2 and R_3 both fail $= F_{23} = F_2 \times F_3$
$= 0{\cdot}20 \times 0{\cdot}20 = 0{\cdot}04$

Therefore:
Reliability of R_2 and $R_3 = R_{23} = (1 - 0{\cdot}04) = 0{\cdot}96$

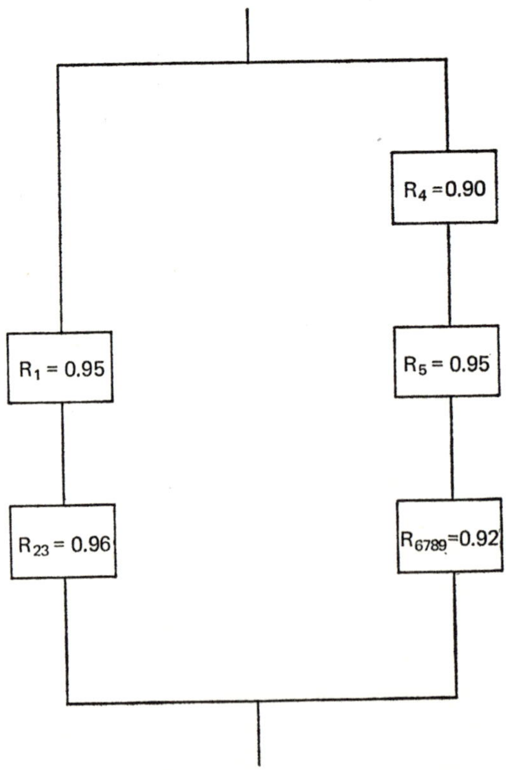

FIG 8.6(c).

4 Similarly for R_{67} and R_{89} we have:

$$F_{67} = (1-0{\cdot}72) = 0{\cdot}28$$
$$F_{89} = (1-0{\cdot}72) = 0{\cdot}28$$

Therefore probability both fail $= 0{\cdot}28 \times 0{\cdot}28 = 0{\cdot}08$

Thus reliability of R_{67} and $R_{89} = R_{6789} = (1-0{\cdot}08) = 0{\cdot}92$

The overall arrangement is now as Fig. 8.6(c).

5 Next we take each series limb in turn.

$$R_{123} = R_1 \times R_{23} = 0{\cdot}95 \times 0{\cdot}96 = 0{\cdot}91$$
$$R_{456789} = 0{\cdot}90 \times 0{\cdot}95 \times 0{\cdot}92 = 0{\cdot}79$$

6 We now have a straightforward parallel configuration.

$$F_{123} = (1-0{\cdot}91) = 0{\cdot}09$$
$$F_{456789} = (1-0{\cdot}79) = 0{\cdot}21$$

Therefore probability whole system fails $= F_s = 0{\cdot}09 \times 0{\cdot}21$
$= 0{\cdot}0189$

Thus reliability of whole system $= R_s = (1-0{\cdot}0189)$
$= 0{\cdot}9811$
$= 0{\cdot}98$ say

8.7 interconnections within series/parallel arrangements

Consider the arrangement shown in Fig. 8.7(a), where stage A is shown in series with stage B, and for increased reliability each has been duplicated. As it stands, the whole will work if either:

R_{A1} and R_{B1} both work, or
R_{A2} and R_{B2} both work.

However there are two more possible combinations which are apparently unacceptable, namely:

R_{A1} and R_{B2}.
R_{A2} and R_{B1}.

We could make both these latter combinations acceptable merely by modifying the configuration to Fig. 8.7(b) in which there is a cross connection in the middle of the network.

We can easily check for a particular case that we have effected

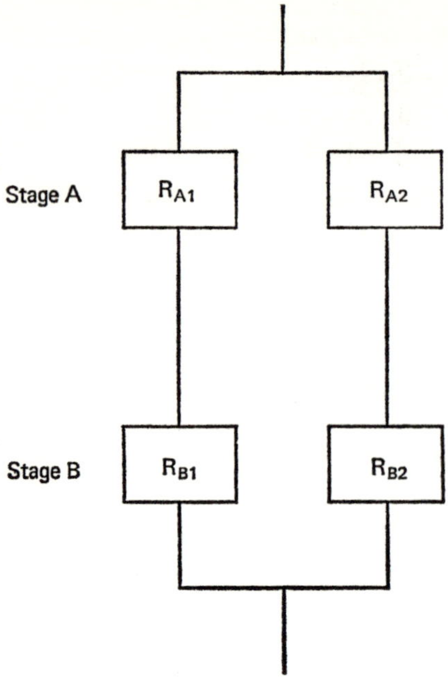

FIG 8.7(a). A parallel arrangement without interconnections

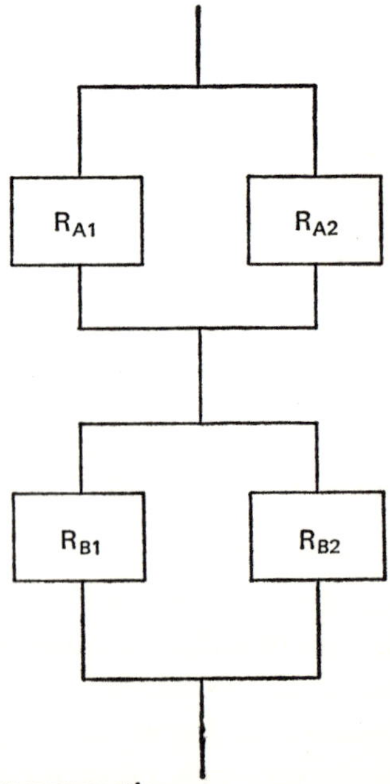

FIG 8.7(b). A parallel arrangement with interconnections

an improvement in reliability. Thus suppose that every unit has a reliability of 0·90:

Without interconnections as Fig. 8.7(a) (1)

Reliability of each branch $= 0{\cdot}9 \times 0{\cdot}9 = 0{\cdot}81$
Probability one branch will fail $= (1 - 0{\cdot}81) = 0{\cdot}19$
Probability both branches will fail $= 0{\cdot}19 \times 0{\cdot}19 = 0{\cdot}036$
Therefore:
Reliability of the whole $= (1 - 0{\cdot}036) = 0{\cdot}964$
$= 0{\cdot}96$ say

With interconnections as Fig. 8.7(b) (2)

Probability one unit fails $= (1 - 0{\cdot}90) = 0{\cdot}10$
Probability R_{A1} and R_{A2} both fail $= (0{\cdot}10)^2 = 0{\cdot}01$
Therefore:
Reliability of R_{A1} and $R_{A2} = (1 - 0{\cdot}01) = 0{\cdot}99$
Reliability of R_{B1} and R_{B2} also $= 0{\cdot}99$
Therefore:
Reliability of the whole $= (0{\cdot}99)^2 = 0{\cdot}9801$
$= 0{\cdot}98$ say

Thus (2) is better than (1) for the same units, in the same application. Although we have considered only two units in series, and two parallel branches, the principle does of course apply to any number.

8.8. standby redundancy

When standby redundancy is used, only one path is energized at a time and the remainder are idle although ready to be brought into action should a failure of the first demand it. This was the case with the standby motor generator in a hospital operating theatre which was mentioned in section 8.1. At first sight it is a more attractive proposition than active redundancy, because the spare units remain new and unused until required instead of wearing out all the time the equipment is operating. We can set out this supposed advantage in terms of the MTBF.

For 1 path only, $\text{MTBF} = \frac{1}{\lambda} = \theta$

For 2 paths in active redundancy, $\text{MTBF} = \frac{1}{\lambda} + \frac{1}{2\lambda} = \frac{3}{2}\theta$

from (8.5a)

For 2 paths in standby redundancy, $\text{MTBF} = \frac{1}{\lambda} + \frac{1}{\lambda} = 2\theta$

(The last statement is true because the second mean life does not start until the first has finished.)

Before we come to any conclusions about the merits of standby redundancy however, we must consider the following facts.

1 The standby unit is not carefully packed away in a specially designed storeroom. Usually it is out on the job, alongside the unit which is operating. According to the application, it may for example be shaken around in a fighting vehicle, smothered with dirt and grease or entombed in a railway locomotive. Thus our assumption that it will remain in perfect working order, ready for immediate use, may be quite incorrect. It may well have an appreciable failure rate while waiting to be used, and the longer it remains on standby duty the greater the risk that it will not work should it be required.

2 For many systems, particularly in the electronics field, the moment of switch-on is particularly hazardous. Stray surges are liable to wander round the circuits, and may well damage any part which is already a bit doubtful. With active redundancy, however, the second circuit is switched on with the first and there is time to check that it works before it is required.

3 Some parts deteriorate if left unloaded for long periods. Large electric furnaces are best kept at least partially energized, since heater failures often occur if for any reason they have to be completely switched off and allowed to cool.

4 It may be possible for two units in active redundancy to share the work, so that each only has to work at half load. This effec-

tively derates both of them, and so may lead to a longer working life as we discussed in section 6.9.

5 Finally consider the procedure when the main unit fails. In only a minority of cases will the sequence be that the operator notices the failure and starts up the standby unit. Usually it will be necessary to install further devices to :

a Detect that the main unit has failed.
b Switch in the standby.

If either a or b above do not work correctly, the standby unit will not come into operation, even though it is in perfect working order.

The detection device will normally be energized whenever the main unit is working, and there are two obvious ways in which it can fail :

a It can fail to detect that the main unit has ceased to function.
b It can cause the standby unit to switch in when the main unit is still working correctly. Although this is a failure of the detection device, it will not of course cause a complete system failure unless the standby unit fails to work.

The switchover device will not normally be energized, but it must function correctly when required. It can also fail in two ways :

a It can fail to switch in the standby unit when required.
b It can switch in the standby unit when it is unnecessary.

Clearly in choosing between active and standby redundancy, we shall have to consider each individual case on its merits, but much often hinges on two factors :

1 The reliability of the detection and switching devices of the standby unit. Remember that we use redundancy because we are not satisfied with the reliability of one working unit on its own. Unless therefore the detection and switching devices are appreciably more reliable than the main work-

ing unit, we may be introducing as much unreliability into our system as we are removing by having the standby unit.

2 It is a fundamental assumption in using a standby unit that its failure rate when shut down, but probably exposed to normal operating conditions, is appreciably lower than if it were energized under the same conditions. If this is not so, then active redundancy may well be a better proposition.

8.9. partial redundancy

So far we have assumed that our system will continue to work provided that at least one path is still in operation. There will however be many cases where, although some failures are permitted, more than one path must continue to work if the system as a whole is to work, and this is called *partial redundancy*. The following are examples:

1 The roadway of a suspension bridge is hung from the chains by a considerable number of vertical members. It might be possible for a few of these verticals to break, without endangering the bridge, but clearly this could not go on until only one was left.

2 An aircraft with four engines could almost certainly land safely if three were still working, but this might be very difficult if only one survived.

Calculation of the reliability, when partial redundancy is used, follows the same pattern as for active redundancy but tends to be more complex since:

1 There are now a number of configurations which lead to failure. Thus if there are 4 parallel paths, at least 2 of which must work, then failure will occur:

a If all 4 fail.

b If any combination of 3 fail.

2 There may be additional restrictions concerning which units may fail. Thus, in our suspension bridge example, not only must a minimum number of verticals survive but they must be reasonably distributed. They could not be all at one end.

CHAPTER 9 ENVIRONMENTAL PROBLEMS

9.1. external and internal environment

We have so far given little consideration to the environment in which our system must work. We can divide this into two groups:

1 *The external environment* by which we mean factors such as weather, mud, sea water, etc. which are external to our system.

2 *The internal environment* which is the effect which parts, etc. have upon each other. Thus one part may be in an unduly hot situation solely due to the heat dissipated by another part in the system.

Every application is likely to present its own environmental problems, but we can nevertheless discuss them in general terms.

9.2. climatic problems

Any system which operates outdoors will experience the full effects of the weather in its locality. Where the system is mobile, as with a fighting vehicle, then it could conceivably be located almost anywhere in the world. We may of course stipulate where it is intended to be used, but we cannot be sure that users will observe our recommendations.

1 TEMPERATURE VARIATIONS

In polar regions the temperature may fall to −40°C, whereas in hot desert regions it may rise to 55°C. Even in one place, day to day and seasonal variations may be considerable, and further differences may be experienced if the system does not remain at sea level, as in the case of an aircraft.

2 PRECIPITATION AND HUMIDITY

Outdoor systems will need protection from rain, hail, sleet and snow. Humidity can be particularly troublesome because it tends to penetrate inside, causing condensation. For a given moisture content, the humidity rises as the temperature falls, and vice versa. Hence any place where the temperature falls sharply at night is liable to have condensation problems. It does not necessarily have to be a wet climate.

In cold climates, build up of snow and ice may obviously be a problem especially on items such as air filters.

3 WIND AND ATMOSPHERIC PRESSURE

It is always difficult to decide what wind speed to anticipate, and there have been one or two structural failures of buildings because the designer did not foresee the strength of wind which was subsequently experienced.

With aircraft and installations at high altitude, allowance must be made for the drop in atmospheric pressure.

9.3. problems arising from the terrain

1 VIBRATION, SHOCK, SUDDEN ACCELERATION, ETC.

Any system which travels is going to be subjected to vibration and shock, and it is almost impossible to foresee exactly what might happen. Roads may be poor or the operator may drive the system over the roughest ground. Rigid construction will be necessary to withstand such treatment. The centre of gravity will need to be kept low and resonant frequencies within the system minimized. We shall try to avoid the use of delicate components which cannot stand up to such treatment.

Sometimes vibration, etc. is really internal in origin, coming possibly from the motors of the system itself or from adjoining machinery.

2 MUD, DIRT AND SAND, ETC.

Obviously almost any system which works out of doors on land will suffer from the ingress of earthy materials. Sand is particularly penetrating, but dirt and mud can get a long way in and occasionally rocks may be flung against the vehicle. Animals and birds may give trouble, especially with air intakes as on a jet engine.

3 SEA WATER AND SEA BIRDS

Naval and other sea based systems tend to suffer from the effects of sea water which due to its salt content is particularly corrosive. If the ambient temperature and humidity are high, sea water may encourage the growth of fungus, while sea birds, like land birds, present a hazard to air intakes, etc.

9.4. man made environmental problems

1 CORROSIVE ATMOSPHERES

Many systems are required to operate in corrosive atmospheres. A common example is the equipment in an electroplating shop, but it applies to many chemical processes. The degree of contamination may be very small, and so the rate of corrosion equally small, but the trouble is that it probably goes on day and night, week after week, until eventually what appeared to be a very minor rate of corrosion has a major effect.

2 RADIATION

Radiation of sufficient frequency and intensity can cause trouble, and this may be a significant factor with processes concerned with atomic energy.

9.5. problems with the internal environment

Inside our system, we at least have the advantage that many problems arise from parts which we ourselves have specified, and which we can therefore change if necessary.

1 HEAT

The dissipation of heat is often a serious problem. If we could space parts adequately there would be little difficulty, but today's trend is towards miniaturization. We are constantly asked to get more equipment into less space. Manufacturers of parts have cooperated by making their products progressively smaller. In many cases, however, the power to be dissipated remains much the same and this results in more heat per cubic metre of equipment.

2 VOLTAGE AND CURRENT SURGES

It is not always easy to prevent voltage surges, and although they may be of short duration they may have dangerously high peak values.

9.6. dealing with environmental problems

Although the way in which we minimize the effect of the environment upon the reliability of any particular system must depend on the nature of that system, there are usually three possible lines of action:

1 Try to select parts which:
a Are insensitive to the environment concerned,
b Do not themselves create an environment which is harmful to other parts.
2 Protect the whole system or sub-system from the harmful effects of the environment.
3 Protect individual parts or groups of parts.

9.7. protection against mechanical damage

In general it is better for a system to be designed to withstand inevitable shock and vibration rather than to try to protect it. There may however be difficulty in foreseeing what treatment it will sustain. Hence when a new design is first put into service early failures may occur from something which the designer did not anticipate.

If a frequency of vibration happens to coincide with the resonant frequency of a section of the system, then that section is liable to shake itself to pieces. Careful tests at the development stage are necessary, to confirm that these have all been foreseen and eliminated. Sometimes a system can be successfully mounted on rubber or metallic isolators or on dampers, but care is necessary otherwise the result may be worse rather than better.

Electrical systems may suffer from the flexing and eventual breakage of leads, and from the dislodging or breakage of delicate parts. The effect of vibration and ambient temperature on the latter cannot always be considered in isolation. Thus a soldered joint may stand up satisfactorily to mechanical vibrations on their own. It may equally stand up to an elevated temperature by itself, but put it where it is subjected to both vibration and an elevated temperature and it may fail.

9.8. dealing with excessive heat

Where a system generates heat, we shall have to make provision for dissipating it. Two important ways of doing this are :

1 Air cooling.
2 Liquid cooling.

9.9. air cooling

Since in most cases surplus heat will finally end up in the atmosphere, we naturally consider air cooling as the first possibility. In a simple case the problem may be primarily one of allowing the heat to transfer easily and therefore quickly from the parts concerned into the air. This is called a path of low thermal resistance. Thus semiconductor rectifiers frequently have metal fins attached to the rectifier element itself. Since the metal is a good conductor, heat transfers quickly from rectifier to fin, and then the large surface area of the fin enables the heat to escape into the air, where it is carried away by convection currents. Silicon and germanium rectifiers are sometimes bolted to a large mass of metal, called a heat sink. This ensures that the heat is removed very quickly from the rectifier itself, but its subsequent loss from the heat sink may still be by air convection.

We can increase the rate at which air removes the heat by means of forced air cooling. There are difficulties however. Thus although we may provide a good current of air over the parts, there are always liable to be pockets which the air stream does not reach, and where the temperature rises excessively. Again, forced air cooling implies the addition of a fan, etc., and this puts up cost, weight and volume. It also reduces reliability, because there is now something else to go wrong. Indeed if the system can only operate satisfactorily when the fan is running, then we must consider the fan as a unit in series with the rest when we calculate the overall reliability, as discussed in section 7.2. Perhaps the most attractive case of forced air cooling occurs on a vehicle or aircraft, which moves through the air rather than requiring air to be forced through it.

9.10. liquid cooling

Air has a low specific heat, and so if there is a great concentration of heat we may have to consider liquid cooling. Water immediately comes to mind as being cheap, plentiful and having a high specific heat. It is particularly attractive for mechanical applications, but its poor electrical properties make it much less suitable for electrical equipment. It is not completely excluded, but it is necessary to isolate the cooling water from the electrical apparatus. The reliability of this isolation must be high, because in most cases it would be disastrous for the water to get into the electrics.

Because of the difficulties which water presents other liquids are commonly used. It is relatively easy to list the properties we want, and although it is not so easy to find one liquid which is good on all counts it is usually possible to find one which satisfies a given application. In particular the fluorocarbon liquids are useful as coolants.

The desirable properties for a cooling liquid are as follows:

1 Electrical properties

a High electrical resistance.

b High electrical breakdown strength.

c Low dielectric losses.

2 Heat transfer properties

a High specific heat.

b Low viscosity, so that it flows well round the circuit.

c Low surface tension.

d High coefficient of thermal expansion, in order to encourage convection currents.

3 Inertness

a The liquid must not decompose in use.

b It must be non-corrosive and non-solvent, so that it does not attack the pipes and other apparatus through which it flows.

c It must not be toxic or inflammable.

d It should have a low freezing point.

9.11. protection from damp and dirt

Some sort of overall protection from the effects of weather, dirt, etc. will often be the easiest approach. In designing suitable casing, however, we may find ourselves at odds with the requirements for cooling. Thus that which effectively keeps water and dirt out also keeps heat in. Various compromises are possible, such as louvres and air filters, the latter being particularly useful when forced air cooling is used.

Individual parts, or small groups of parts, may be protected by encapsulation in a solid material, usually a thermosetting resin. If well done, the resulting protection is very effective, but difficulties may arise if the part concerned is damaged or restricted in some way by the presence of the resin. Thus :

1 Some resins contract slightly on setting, and if by then they have bonded to the surface of the part, the stresses may damage it.
2 Water may get in wherever the bonding is not perfect and, if it is difficult for the water to get in, it will be equally difficult to get it out again.
3 Thermal expansion may cause problems.
4 In a few cases, corrosion may take place.

It is probably best for the part manufacturer to do the encapsulation, if he can, since he will know best what is likely to upset it.

CHAPTER 10 DESIGN FOR EASE OF OPERATION

10.1. the competence of the operator

In designing equipment for other people to use, we must consider how skilled or otherwise they are likely to be. Almost certainly they will not be equal to the development engineers who will test out the design. Indeed in some cases, as when the equipment is to be exported to underdeveloped countries, it may be that only relatively unskilled people will be available.

The system should be designed to be as easy as possible to operate correctly, and as difficult as possible to operate incorrectly. If two controls should never be operated simultaneously then it helps to interlock them, so that the operation of either puts the other out of action. Alternatively, but second best, the controls may be spaced so far apart that one operator cannot reach both at the same time. Unfortunately, however, the presence of a second operator can beat this safeguard.

Since, sooner or later, perhaps due to illness or strikes, an untrained operator may try his hand it is essential to devise the equipment so that no injury or damage is caused by any mistakes which are made.

10.2. the effect of fatigue and working conditions

The more tiring an equipment is to operate, the greater the risk of a mistake and the shorter the period for which an operator can work without a rest. It has been said that an ordinary machine shop lathe is really suited to an operator about 4 feet tall, with an arm span of 8 to 10 feet! Clearly we must not expect our operator to be a contortionist.

A lot can often be done quite easily to take the physical effort out of a job. Thus :

1 Constant bending, either to pick up materials or to operate very low controls, is unnecessarily tiring.
2 Heavy weights should be moved mechanically rather than by the operator's own exertion. We are here in the realm of method study and the reader who is interested may care to consult one of the many books on the subject, such as *Work Study* by R. M. Currie, published by Pitman.
3 If the equipment generates excessive heat, it is in principle better to design so that it is not dissipated in the direction of the operator, rather than to leave the customer to provide protective clothing. The latter is usually uncomfortable, and so may not always be worn.

10.3. the layout of controls and tools

Where tools must be used in a particular order, it is a great help if they are presented to the operator in that order. Controls too should be arranged logically. The keys of a piano are arranged in sequence from the lowest note to the highest, and no one would have them otherwise. Yet the controls on panels are sometimes placed quite illogically, perhaps to suit the arrangement of the equipment behind the panel. As a result the operator has to learn to operate them in some arbitrary order. A particularly inexcusable example of this occurs where the illogical order is created by the designer, in order to lay out the panel so that the controls are symmetrical about some centre line. Mistakes are then likely, especially under pressure. A considerable variety of knob shapes is now available, and it helps if all the controls in a particular group are different so that the operator knows by feel which he is holding.

Controls should be labelled as simply as possible, but with enough information so that the operator understands at once what they do, without consulting the instruction manual to find out.

10.4. indicators

By indicators we mean dials, gauges, lamps, bells, etc. which will show the working state of the equipment. The majority are either visual or aural. Thus a light may come on or a bell may ring to warn the operator that a particular cycle of manufacture is complete.

Lamps in general can only be either lit or extinguished, and so can only indicate one of two alternatives, usually on or off, whereas meters can give an actual reading. We shall decide each case on its merits but, broadly speaking, if the operator only wants to know whether a particular part of a system is on or off then a light is usually preferable. It is not only cheaper but easier to observe. One can see at a glance whether a light is on, even from across the room, but the needle of a meter must be read. Since however a meter conveys more information it must be specified wherever an operator needs it.

Sometimes a lamp is all that is necessary for routine operation, but a meter is required for setting the circuit conditions and for fault finding after a failure has occurred. Probably in most cases it is easier to settle for a meter, with the correct needle reading clearly marked in, say, red, although sometimes it will be more convenient and cheaper to include the meter in the checkout equipment, which may be common to several identical systems.

10.5. the arrangement of meter dials on the display panel

Meter dials are easiest to read if they are roughly at eye level. When they are very high or low they are not only tiring, but there will inevitably be errors due to parallax. Avoid highly reflective surfaces, such as chromed meter rims or shiny instrument panels, since these may dazzle the operator.

Ensure that meters are calibrated directly in the units to be observed. Thus suppose the meter indicates the moisture content of timber being fed into a process. The meter may in reality be a micro-ammeter, but it should under no circumstances be calibrated in microamps with a chart to convert to percentage moisture. This is not only time consuming but also leads to errors, and anyway sooner or later the chart gets lost. It must be calibrated directly in percentage moisture.

A dial should be calibrated so that it can be easily read to the required accuracy, but not to an accuracy which exceeds that of the meter. Thus it is misleading for an operator to quote a reading of say 2·34 V, when the meter itself is only accurate to 0·5 V.

Nowadays digital type meters, which display actual numbers, are becoming increasingly available. Other things being equal they are preferable, because they are quicker to read and less prone to operator error than the traditional analogue type meter. Fig. 10.5(a) illustrates this point.

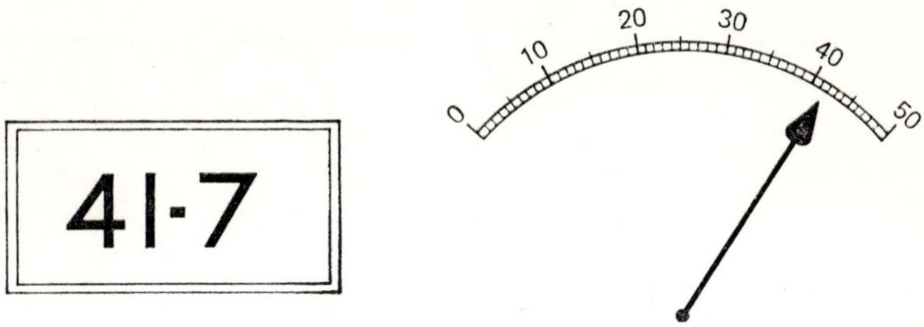

(i) Digital type meter

(ii) Analogue type meter

FIG 10.5(a). i Digital type meter   ii Ordinary analogue type meter

Where a group of meters must be inspected periodically to see that their reading is correct, their arrangement on the panel is important. Fig. 10.5(b) (i) shows a group of 6 meters, and in order to check that each displays the correct reading each must be individually read.

In Fig. 10.5(b) (ii) the meters are arranged in a row, and mounted so that the needles are horizontal when they show the correct reading. A mere glance is enough to ascertain that the third meter from the left is showing an incorrect reading. The same principle can be applied where it is more convenient to mount the meters vertically, as in Fig. 10.5(b) (iii). Now it is best for the needles to be vertical when they are correct. It is not necessary to have the meters in long horizontal or vertical lines. Fig. 10.5(b) (iv) shows how we can return to two rows of three, and still retain the advantages discussed.

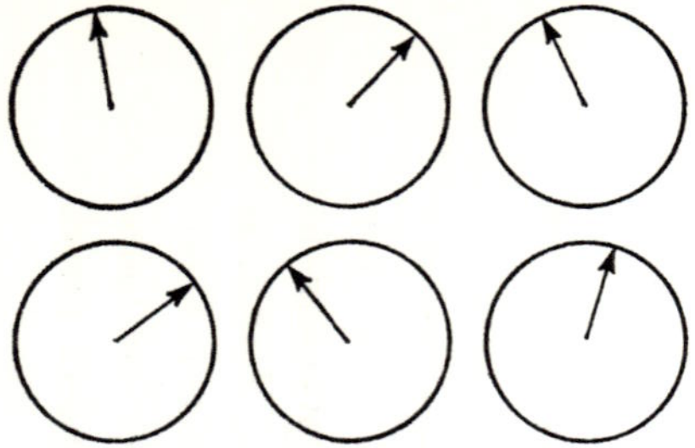

(i)

(ii)

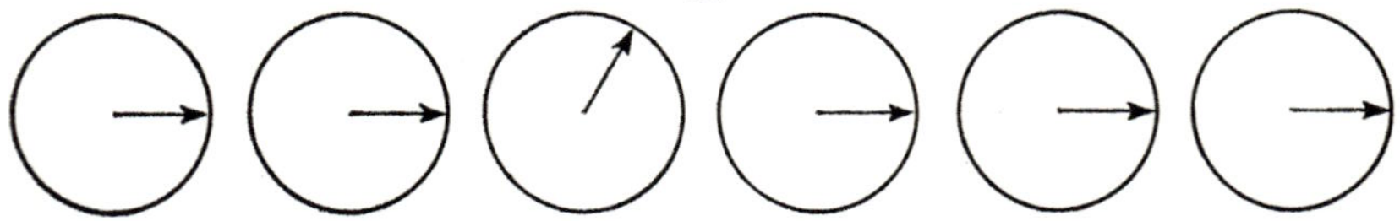

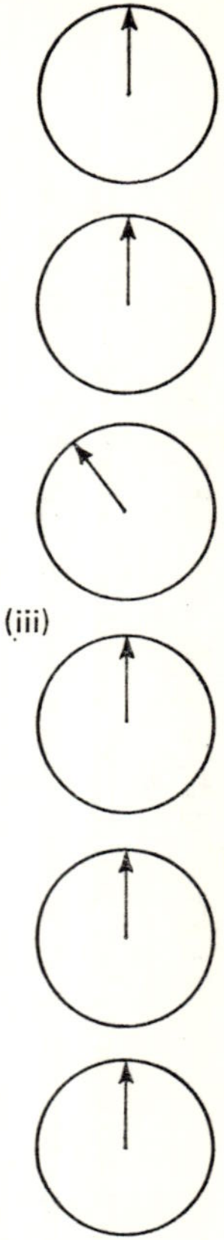

(iii)

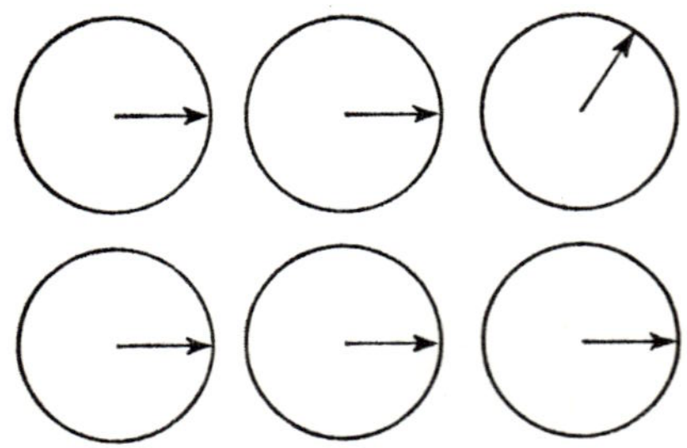

(iv)

We must make sure that if the power fails the needles will not come to rest in any valid position, which the operator might mistake for an operational reading.

10.6. electrical controls and responses

Frequently a control is related to a particular meter, so that if, say, the voltage shown is low, the operator turns up the knob which controls it. In such cases:

1 The knob should turn the same way as the needle, so that turning it clockwise causes the needle to go clockwise.

2 The knob and the dial should be placed as close together as possible, and there should be no doubt which knob controls which dial. In particular the control should not be round the back of the equipment, when the dial is in front. This situation is sometimes found where setting adjustments are concerned and the maintenance man, having carried out a repair, must adjust the equipment by altering a control at the back then walking round to the front to read its effect on the meter.

3 There should not be too much delay between the alteration of the control and the response of the meter. If the operator turns up a control, and apparently nothing happens, he is likely to turn it up a bit more. When the response comes, the needle will then overshoot its intended setting and the operator will have to turn the control down again, probably this time undershooting and so on. This phenomenon is called hunting, and it occurs on some automatic as well as manual controls. Thus the thermostat on a furnace will probably cut out too late to prevent an overshoot and then cut back in too late to prevent an undershoot. There are methods by which hunting can be greatly diminished, usually by beginning to reduce the input power well before the cut out temperature is reached, and then leaving a small amount of power still on after cut out.

10.7. mechanical controls and responses

The principles which we have discussed for electrical systems apply equally to mechanical equipment, although they may be more difficult to observe. Thus :

1 Wheels and levers should move logically with respect to the function they control.
2 Controls should be placed where the operator can see at once the effect of operating them. He should not have to walk round to see whether something which he has started up is in fact running.
3 Response to the operation of controls should be quick enough to avoid hunting, although this may not always be easy to arrange. Thus it may be possible to shut quickly the gate controlling the input of an ingredient to a chemical process, but there may be some delay before the chute has cleared of material which was already past the gate.
4 Try to avoid controls which require a considerable physical effort to operate them.

Increasingly nowadays we find that the controls of mechanical equipment tend to be at least in part electrical, and then all that we said in section 10.6 applies.

10.8. alarm signals

In both electrical and mechanical equipment we should include alarms to warn the operator when all is not well. An everyday example is the red light on the dashboard of a car which comes on to warn us that the battery is not charging. Warning lights vary in their degree of urgency. Some of the lights we discussed in section 10.4 might be regarded as warning or advising the operator that a particular part of the circuit is or is not switched on.

In this section, however, we are considering the more serious case, where something has positively gone wrong. Our first consideration when this happens must be to see that the system will fail safe, as we discussed in section 7.11. It may well be better to ensure that the whole system will automatically shut down rather than to warn the operator, who must then decide to shut it down and act upon his decision. This is particularly important with manufacturing processes, where delay in shutdown may result in thousands of pounds' worth of materials ending up on the scrap heap.

We shall of course require an alarm, as well as an automatic shutdown, because once the plant has stopped adjustments and repairs must be immediately put in hand to get it working again. If warning lights are used, they will need to be larger than those for routine indication. Bells and buzzers, particularly if they have a penetrating note, are even better. They must positively crave attention.

It is important to make sure that the meaning of an alarm is clearly shown on the control panel. In an emergency, it is no good if the operator then has to start a panic hunt for his instruction manual to find out what the alarm means.

The inclusion of alarms and cut outs inevitably means something else which could go wrong. An unreliable cut out might be worse than no cut out at all. Since it would not be used every day the operator and maintenance man might not realize that it it was not in working order, until an emergency arose, and by then it would be too late.

10.9. ergonomics

The science of fitting the job to the operator's natural abilities, rather than assuming that the operator will somehow cope with a badly designed equipment, is called ergonomics or sometimes human engineering. This subject can be pursued in one of the standard textbooks, such as *Ergonomics—Man in His Working Environment*, by K. F. H. Murrell published by Chapman & Hall.

10.10. imaginary malfunctions

Sometimes when a system does not appear to work properly, it is merely because it has not been correctly operated. The operator may not realize this, however, and proceed to shut it down and call the maintenance engineer. Hence good operating time is lost and the maintenance engineer's time is wasted, and all because the operator's skill is not quite equal to the demands of the system. We can help by designing the system so that it is as easy as possible to operate.

CHAPTER 11 DESIGN FOR EASE OF MAINTENANCE

11.1. the importance of availability

When a breakdown occurs, our customer's chief interest will be to get the system back into service with a minimum of delay and expense. Indeed he is more likely to be concerned with its availability than with its reliability in the strict sense. Hence we must consider how to make our system quick and easy to service.

11.2. fault location routine

A set routine for fault location is essential otherwise a lot of time can be wasted, as the following illustration shows. A particular piece of electronic equipment failed from time to time and the repair technician was duly summoned. He had then to decide which part was faulty. Without a proper routine, however, this was very difficult, and so he would contemplate the circuits until, as it seemed, he took a 'dislike' to a particular part, usually because it was dirty or shabby looking rather than because it was faulty. The part concerned was replaced, but as it was unlikely to be the one which had failed this seldom cured the trouble. Hence the process was repeated and sometimes after 2 or 3 attempts the fault would be located, but even so the system did not always work satisfactorily because by then the chances were quite high that the attempted repairs would themselves have introduced one or more additional faults. Hence a more experienced engineer had to be called to check the whole system through and get it working again. It was no good blaming the technicians for this. They did their best, but they needed a routine devised by someone familiar with the function of every part of the system.

The designer and the development engineer might cooperate

to produce a fault finding routine. It should be as simple as possible, and set out on a step by step basis. The approach will probably be first to determine which block in the system contains the fault, then which unit within that block is faulty and so on so that in as few steps as possible one is able to locate the actual part which has failed.

In simple cases the fault finding routine can be expressed as a family tree, but otherwise tabular form may be more convenient. Fig. 11.2(a) below shows how the routine might look for a relatively simple piece of automatic test equipment.

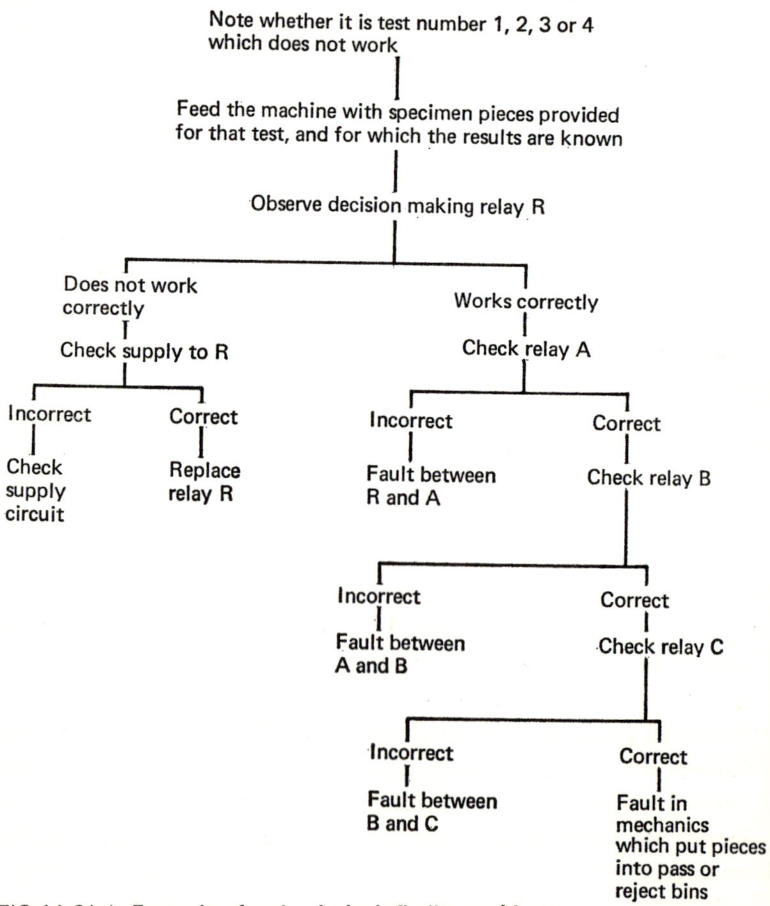

FIG 11.2(a). Example of a simple fault finding routine

Where a mechanical equipment is concerned, it may be difficult even after the fault area has been located to see exactly what is wrong because of the speed with which the mechanism works. In such cases, a stroboscopic lamp may be useful in order to apparently slow the process down. A printing press, for example, works too quickly for the average eye to follow and hence it may be very difficult to observe why a particular fault is occurring. A lamp which flashes just a little more slowly than the rate at which printed copies are being produced will make the operation appear to go very slowly.

11.3. provision for fault location

Having decided roughly what the fault location routine is to be, the designer must consider how he can make it easy to carry out. Possibly the best solution is to arrange for lights and meters on the control panel, to indicate precisely where a fault is. The next possibility is to provide check points, so that the required test equipment can be plugged in. It may also be desirable to provide alternative units which can be plugged in and so allow the maintenance engineer to eliminate sections of the actual system in turn.

A simple number code probably assisted by colours should be provided, so that any part can be located quickly. Remember however that colours fade and get dirty and anyway some people, especially men, are colour blind.

Throughout fault location and repair, there is liable to be a lot of poking about inside and therefore we must have particular regard for safety. In a factory, production machines have guards which are often interlocked so that, if the guard is removed, the machine goes dead. Guards such as the sides of electrical cabinets which are only likely to be removed by the maintenance engineer are not always interlocked in this way. This is curious because it is every bit as serious if a maintenance man gets hurt as if it had occurred to a production operator. Hence we must ensure that checks which must be done with the system alive can be done safely, preferably without removing protective guards, and that once these guards are removed the equipment is safe and preferably dead.

11.4. test equipment

For a simple system, the test equipment may consist of no more than one or two meters and some hand tools. As systems become more complex, however, we find that the test equipment itself becomes complicated and that it is necessary to consider its reliability. Thus very roughly the whole of this book can be applied to the test equipment as though it was an independent system. It should be designed to be as reliable as possible, using techniques like redundancy where necessary, and it should be as easy as possible to check and maintain. Clearly endless confusion can be caused if a system which has itself failed is checked out with test equipment which does not work properly.

The test equipment may need to include provision for :

1 Simulating a load or other condition.
2 Providing particular input voltages, etc. so that the circuit response can be checked.

Since all this will enlarge the test equipment, we then have to consider :

1 Weight.
2 Portability.
3 Robustness.
4 Cost.

Remember that test equipment must usually be mobile, in order to service a number of identical systems.

11.5. fault correction

Once a fault has been located in a system, then repairs and adjustments must be carried out and again the designer can help to make this as easy as possible. Thus :

1 Parts which have high failure rates must be easy to identify and get at.
2 Fixing devices must be easy to release and reassemble. Thus nuts, bolts and screws in awkward places should be avoided.

It is difficult to insert or tighten a screw into a position which one cannot see and perhaps cannot even feel directly with the fingers.

3 Plugs and sockets are attractive, because they make dismantling and reassembly easier, although their own reliability may present a problem.

4 Make sure that visual inspection is as easy as possible, and that parts such as nuts and screws do not when released immediately fall down inside.

5 Where appropriate, make it possible to replace a complete unit, so that the faulty one can be taken away for repair.

11.6. working conditions of the maintenance staff

The designer must try to anticipate the actual conditions under which maintenance may have to be done. Thus :

1 The skill of the maintenance staff may be very limited indeed, and in remote areas there may be no one to whom they can turn for help. This applies particularly to mobile systems such as vehicles and tanks which might be almost anywhere at the moment failure occurs. It must then be possible to carry out essential if temporary repairs so that the vehicle can get back to base.

2 The maintenance equipment available may be similarly limited. Fault location may be easy if the correct instruments are available, and yet become well nigh impossible if they are not. Heavy items such as replacement engines can be handled without difficulty when a hoist, inspection pits, etc. are available, but may otherwise be very difficult.

3 The repair workshop at base probably has reasonable working conditions, but out on the job they may be most unreasonable. Thus the inside of an aircraft standing on a desert airstrip may be intolerably hot. At the other extreme, repairs may have to be done in arctic conditions, in a howling wind and in the dark. We must consider whether the necessary repairs can be done in

full cold weather clothing, and in particular whether the maintenance man can do all that is required with warm gloves on his hands. Even if he decides to take them off, his sense of touch will be blunted by the cold.

4 Consider what spares are likely to be available, remembering that it is much more difficult to obtain special parts than standard ones.

11.7. instruction manuals

A system must have a clear concise instruction manual written in the native language of the user, yet curiously the initial approach of the designer should be almost the opposite of this. Throughout design and development, he must consider not how much information he should cram into the instruction manual, but rather how to minimize its necessity. Thus :

1 The method of operation should as far as possible, be clear without reference to the manual. We discussed this in Chapter 10. Controls should be arranged and labelled so that their use is self-explanatory.
2 The same applies to maintenance. It is much better for adjustment points to be readily visible and accessible, than to have to consult the instruction manual to find out where they are hidden.

Thus our first objective should be to make the manual unnecessary. It is unlikely of course that we shall completely succeed, so thought will have to be given to writing it. Thus :

1 Use clear diagrams wherever they would be useful. A simple sketch often saves a lot of writing, which the operator may not fully understand.
2 Try to put a diagram on the same page as the text to which it relates.
3 Keep the wording concise and simple. In particular avoid technical terms which, while they may be in standard use

in the design department, may mean little to the operating and maintenance staff.

4 Make sure that it is easy to locate any piece of information which may be required. Few people are likely to read the manual from cover to cover, and when a particular part has failed the maintenance engineer wants to know quickly how to locate and replace it, and he does not expect to have to read half the instruction manual to find out.

5 Provide a good cross-referenced index, so that the users can find what they want even though their terminology may be somewhat different from that used by the manufacturer.

11.8. after-sales service

The complexity of modern equipment makes it increasingly impracticable for customers to do more than routine maintenance, and some customers anyway prefer to rely upon the services of the manufacturing company. Hence we must provide an after sales service, whose objectives might be as follows.

1 To offer an efficient maintenance service at reasonable prices. This will include:

a Routine servicing.

b Repairs and adjustments, whether made necessary by fair use or by misuse.

c The sale of spares to customers who prefer to do their own maintenance.

d Dealing courteously and efficiently with all customer complaints, whether or not they are justified.

2 To maintain a continuous relationship with customers, so that when the moment comes to purchase further equipment they instinctively turn to our company for it.

3 The last objective may be to operate profitably, although a good after sales service will probably justify itself, even if it makes a modest loss. Thus its contribution will include:

a Improvement of customer relations.

b Collection of valuable information about failures and other field experiences, for analysis and use by the reliability engineers.

The increasing use of mass production has the effect of cheapening products. There has not however been any corresponding reduction in the cost of making repairs. Hence the tendency for it to be cheaper to replace a faulty unit with a new one rather than to attempt to repair it.

11.9. stock control of spares

The provision of spares and replacement parts presents a particularly complicated problem. We must not only be able to supply spares for models in current production, but also for those which have gone out of production but are still used by customers. Sometimes to please a customer we may have to hand make a part which is no longer available. Invariably, however, such a part is sold at a loss, and if this happens too often it eats into the profitability of the after-sales service department.

A further difficulty arises from the necessity for quick delivery. Repair time to the customer is the period an equipment is out of use, and it is no comfort to be told that repairs can be done in two hours if the necessary spares take two months to obtain. In many cases this will necessitate stocks of spares in various parts of the world, with a competent staff to look after them.

11.10. collaboration between after-sales and the rest of the company

The after-sales service department can offer considerable assistance, especially in quality and reliability, to the rest of the company.

1 DESIGN

The service department should be consulted right from the first conception of a new design. They more than anyone are familiar with the snags and difficulties which customers have experienced with previous models, both in reliability and maintainability.

2 PROTOTYPE

When snags are experienced during prototype tests, the service department will be able to comment upon their seriousness from the customers' point of view. Sometimes prototypes tests may include getting a few potential customers to try out a new design in their homes or factories.

3 CUSTOMER EXPERIENCE

Once the new design is on the market, the service department and the sales staff collaborate to collect and collate customers' experiences with it. Any serious design and manufacturing faults will quickly come to light and in many cases the general pattern will be obvious within a few months.

CHAPTER 12 MANUFACTURE FOR QUALITY AND RELIABILITY

12.1. the need for prototype tests

Each design must be proved before it goes to production and therefore a few prototypes are first produced and thoroughly tested. If the tests reveal any shortcomings in either performance or reliability, the design must be modified and the tests repeated.

In practice, design and initial tests are intermingled. Thus:

1 The design staff find that they are short of information about the performance of some item.
2 Tests are done to fill this gap.
3 Design proceeds until another gap in the designers' knowledge is revealed, when more tests are put in hand.

Nevertheless it is very desirable that, when the design is complete, formal prototype tests should be carried out by competent staff who are independent of the designers. This should be done along the lines discussed in sections 21.4 and 21.5.

When well controlled tests are carried out on prototypes which are themselves a faithful interpretation of the design, the observed reliability is called *the inherent reliability*. It represents the maximum that that design is likely to achieve under those conditions. Later, when the design is in production, there may be some loss in reliability if the factory fail to manufacture exactly to the design specification although good quality control will reduce this risk to a minimum.

A further loss of reliability may occur if the customer treats the product worse than has been anticipated by the test conditions. Strangely, however, a gain in reliability is sometimes observed in customer service. This can happen if the manufacturing company set their test conditions to match the worst use conditions a customer might impose. Since many customers will not demand the worst, the average performance in service may be better than the test results suggest.

12.2. the quality standard

Once the design has been proved, drawings and specifications can be sent to the factory. In principle, these should show exactly what *quality standard* is required though in practice such perfection is never achieved. Therefore the quality and production engineers must first make sure that they are quite clear about the standard demanded. They will, we hope, have been consulted throughout the design and prototype stages in order to ensure that the final design is a thoroughly practical production proposition.

All the qualities and functional tests which are required can be expressed as either *variables* or *attributes*.

1 *Variables* are qualities which are observed in figures.

For example. A diameter = 25·0±0·03 mm

An output voltage = 150 ±0·2 V

Variables are preferred to attributes because their meaning is precise. Both the millimetre and the volt are accurately defined units, so there can be no doubt about what the above means. However no measuring instrument is ever absolutely accurate and precise, and care is necessary to ensure that the errors of measurement are small compared with the tolerances to be observed.

2 *Attributes* are qualities which will not be observed in figures. They divide into those where it is merely inconvenient to use figures and those where there is some practical difficulty.

The use of a go/no go gauge in a machine shop is typical of where a quality could be observed in figures using a micrometer, but where a gauge is often both quicker and easier. Since the gauge can be precisely set in terms of variables there is no doubt about the quality standard.

It is the qualities which cannot easily be related to figures, which give rise to difficulty. For instance:

This casting must be free of porosity.
This gear box must run quietly.

Opinion immediately enters into quality standards of this sort. What one person thinks is a quiet gear box someone else will consider much too noisy. A lot will depend upon where it is to be installed. If it is going in an already noisy factory probably no one will notice it, but if it will be in a quiet situation it will seem to make a dreadful noise. Noise levels can of course be measured in figures, and indeed there are probably very few qualities which could not be observed in figures if the cost of doing so was justified.

For qualities which are not to be measured in figures, the quality engineer must set some sort of practical standard. Thus he might select an X-ray negative showing the maximum amount of porosity in castings which is to be considered as just acceptable.

It is important for the designer to realize that if he fails to specify precisely what quality he requires, someone in the factory will have to do it for him. If the quality engineers or their equivalent do not do it, then the shop floor operators will be left to give their own interpretations. If these do not match up to what is required, then the designer has only himself to blame. Everyone else has done their best.

12.3. planning to achieve the required quality

Once the quality standard has been specified, the factory personnel must plan to achieve it. This is often long and involved but unless it is done properly there will be endless trouble when production starts. We shall confine ourselves to the following aspects as they affect reliability:

1 The method of manufacture.
2 Production equipment.
3 Inspection and test equipment.
4 Supplies of raw materials and parts, etc.

12.4. the method of manufacture

The factory engineers must devise methods of manufacture which will achieve the required quality and quantity as economically as possible. It should be as easy as possible to do each job correctly, and as difficult as possible to do it incorrectly. The designer can help by ensuring that as far as possible his designs will not assemble incorrectly. Equally the methods engineers will try to ensure that jigs, etc. will only accept parts if they are correctly positioned.

12.5. the production equipment

The production engineers must provide suitable plant and equipment to achieve the required quality. Where high quality is demanded a machine capable of achieving it must be specified, but where only low quality is required it is inefficient to use an unnecessarily accurate and precise machine.

12.6. inspection and test

The quality engineer must provide the necessary inspection and test equipment and, as with methods, consideration starts at the design stage. Whenever a designer calls for a particular quality there is an implied assumption that a practical and economically acceptable method of verifying it exists. If this is not so, then the quality has no practical meaning. At every stage the designer must ask himself not only what quality he wants, but how that quality will be checked when it is in production. This may necessitate the purchase by the quality engineer of specialized test equipment which he may have to order long before the final manufacturing specification reaches the factory. Hence the need for close cooperation between the designer and the quality engineer.

The quality engineer must plan the shop floor quality control, to ensure that as far as possible everything is made right first time. The following are some of the techniques open to him.

1 FIRST-OFF INSPECTION This is a technique for ensuring that every production run starts correctly and that the set-up of machine, jigs and fixtures is satisfactory.

2 OPERATOR CONTROL In this method, the operator is given the necessary inspection equipment so that he can check his own work as he goes along. Its great advantages are that:

a It encourages the operator to take an interest in his quality.

b If properly worked, it ensures that almost all the items made are satisfactory, because the operator knows at once if his quality goes wrong.

3 PATROL INSPECTION Periodically an inspector checks a sample of each operator's work to make sure that everything is still all right. This ensures that there is never more than one period's work which is out of limit before the fault is discovered. Patrol inspection is used to complement and back up operator control; it should not be used to duplicate it. First-off inspection can be used with either operator control or patrol inspection.

12.7. the provision of satisfactory materials and parts

No one can make a good job out of unsatisfactory materials. Again consideration starts at the design stage, where an adequate specification must be provided for each material and part required. It does not follow that the designer will actually prepare them, since many standard specifications exist, and part manufacturers are usually prepared to specify the important features of their own products.

It is important to realize that the raw materials, parts and other items used by one factory are in fact the finished products of the suppliers' factories. Thus we are really specifying the final quality standard which our suppliers' quality control must achieve. Clearly our demands must be reasonable in relation to the price we are prepared to pay.

Although the designer must specify exactly what is required he should not tie his own factory down more than is necessary.

For example, he should try to avoid specifying one particular make with no alternative, since this prevents the factory purchasing department from shopping around for the best value.

Sometimes it will be necessary to carry out tests on samples from a potential source of supply. Where these are tests for quality, they will probably not take too long. To check reliability, however, it will probably be necessary to put a suitable sample on life test. As we saw in section 6.7, part failure rates are usually very low, and we may have to test a lot of parts for a long time to get any useful results. Hence it is important that such tests are put in hand as early as possible. We can then advise the purchase department what to buy and where to get it.

Having agreed the specification and paid a fair price the factory have a right to expect that the materials they receive from the supplier will conform to specification. Unfortunately this is by no means certain, and most companies find it necessary to employ some form of goods inwards inspection. Satisfactory batches are then retained and unsatisfactory batches are, as far as possible, returned to the supplier.

Inspection results should be recorded for future use. We can do this by providing a simple record card for each material or item purchased, using separate cards where the same item is bought from more than one supplier. As information builds up, we can determine which are our best suppliers and which items give trouble regardless of the supplier. This enables us to introduce some form of *supplier grading* or *vendor rating*, so that:

1 We can prepare a list of preferred suppliers, as discussed in section 6.8, and give copies to:

a The designers.
b The purchasing department.

Where suppliers are found to be reliable, we can reduce the amount of testing we do on goods they send us.

2 Unsatisfactory suppliers can be discontinued or investigated. If we are obliged to buy from an unreliable supplier, because

there is no one else who can supply the item concerned, we can at least protect ourselves with an adequate amount of acceptance testing.

Reliability acceptance tests are discussed in detail in Chapter 19.

12.8. manufacture to specification

As the previous sections show, a great deal of quality planning is necessary before manufacture can begin, and the better it is done, the less risk of snags on the shop floor. Indeed it is when production begins that any shortcomings in the planning are discovered. There are sure to be a few operator errors as well, but we must be careful not to blame the operator for planning errors just because they are discovered at the shop floor stage.

The quality engineer will have planned the routine by which the factory will keep the product quality correct at every stage of manufacture, and this may well include various quality control charts. Basically these are graphs which plot the quality produced against time. Various control limits are added to show how much a quality may vary before the process is to be stopped and reset.

From our discussion in section 6.10, it follows that most qualities should have a correctly centred Normal frequency distribution curve, with as low a variability as possible. Thus suppose we have to make a batch of resistors with a nominal resistance of 50 Ω. Of the curves shown in Fig. 12.9(a), we shall aim for curve A, since curve B has a larger variability and curve C has a mean value of 49 instead of 50 Ω.

Whenever we find a frequency distribution curve which is not Normal we shall investigate the reason, because this is an indication of unnecessary variability. Our immediate objective will be to reduce the proportion of out-of-limit work produced and so to reduce the cost of scrap and rework. Where however the qualities will affect subsequent performance, there is a second advantage which we illustrated in Fig. 6.10(a). Reduction in variability actually increases the safety margin and therefore reliability of

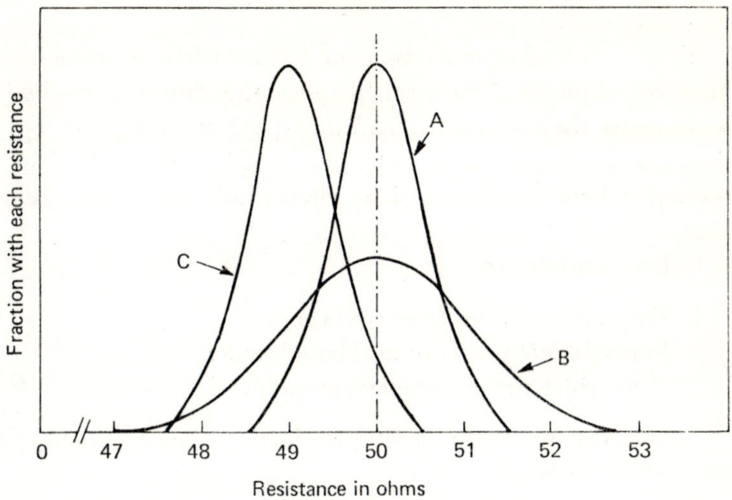

FIG 12.9(a). Frequency distribution curves of processes with different means and different variabilities

the items concerned without any increase in their average quality level.

The quality engineer can estimate the capabilities of all his processes, and he should feed this information back to the designer so that the latter can set his tolerances intelligently, taking advantage of good processes which enable him to set tighter limits and giving a bit back where the process capability is not so good.

The reliability minded quality engineer will keep a close look out for things which, while they do not affect the quality immediately, may cause failure in the customer's hands. Dirt has proved particularly troublesome in this respect. For example small pieces of dirt or metal filings can easily get left in an air valve during assembly, but unless they get on to the value seat it will still pass final test. Some time later, however, the dirt may work its way into the seat and the valve will fail to close properly. Notice the close connection between quality and reliability. If the dirt is already in the seat during final test in the factory it will be rejected as a quality failure. If however that same piece of dirt works its way into the seat during customer service it will count as a reliability failure.

Similar difficulties may be experienced with electrical assemblies. Small pieces of dirt or metal get inside during assembly but, so long as they occupy a harmless position, they will not be detected by factory tests. Later the metal works its way into a position where it causes a short circuit and the whole assembly fails.

Other examples are:

1 Trapped leads and loose connections.
2 Joints soldered with an acid based solder.
3 Parts which are not properly secured.

Any of the above can easily pass factory tests and then fail in the customer's hands.

Basically then the factory's responsibility for reliability involves operating an efficient quality control system, but includes efforts to reduce the inherent variability of all manufacturing operations and attempts to foresee everything which, while all right now, could cause failure later.

12.9. final test and inspection

For all but the simplest products, it is common for the whole factory output to be submitted to a final test before it is despatched to the customer. In principle this is done to protect the company's good name. If anything should have gone wrong during manufacture, and yet remained undetected, the final test offers a last chance to find it and so avoid customer complaints.

In addition, however, it provides the quality engineer with a lot of information. As we shall see in section 17.9, he can analyse faults and rejects, to discover which of them cause most of the trouble, and he can then take action to reduce these faults in the future. The final test, however, should not be regarded as part of the quality control system itself; it is much too late for this. This is merely a check or quality audit, to make sure that quality control in its widest sense is working satisfactorily.

Even when every item made is tested, the actual number of tests applied to each may be quite limited. In a simple case it may be no more than a functional test to see that it works, and a visual check to see that it looks all right. Complicated and expensive items of equipment will probably be subjected to more tests, but it is usually almost impossible to cover everything. Thus consider as an example the modern car. If we listed everything which could be wrong, we should require thousands of checks. We cannot possibly check everything on every car, and anyway we hope that shop floor quality control will have taken care of them as production proceeded.

As an additional safeguard, however, we may decide to take a small sample of the cars produced and subject them to a really searching examination. Such tests should be carried out by staff who are quite independent of production, and they should be instructed to look at the car from the customers' point of view rather than that of the car manufacturer. Thus wheel alignment, brakes, steering, engine performance, body finish and its ability to keep out heavy rain can all be minutely examined. The object of such tests is to try to look into the future and to predict whether the equipment concerned is likely to perform satisfactorily for as long as the customer expects.

12.10. quality and reliability control in service organizations

The principles of quality control apply, with a slight shift of emphasis, to service organizations. Indeed what in the factory might count as quality failure could in a service organization count as a reliability failure. The steps set out in sections 12.2 to 12.8 broadly apply. Thus :

1 SET THE OBJECTIVE A service organization must start by setting out exactly what service it intends to provide. For instance :

a In a hospital, it may be a particular course of treatment.
b In the armed forces, it might be that a particular weapon

must always be able to give a specified performance, on demand.

Notice the reliability content. Whether the weapon performs as intended is a question of reliability. The same would apply in a hospital to the performance of, say, an iron lung.

2 PLAN TO ACHIEVE THE OBJECTIVE

a The precise method must be set.
b Suitable equipment must be obtained.
c Suitable materials must be provided, and there must be controls to ensure that they are of adequate quality.
d Competent staff must be obtained and trained.
e A control system must be devised to ensure that the whole of the plan is carried out as intended.

3 CARRY OUT THE PLAN We now do what we have planned. If our planning and control are good, we shall achieve our objective. If for any reason we fall short, then we must correct the plan for next time. It is bad enough to fail once; we cannot allow it to happen a second time. In so many service type activities lives are at stake, and our reliability must approach certainty. This includes not only the reliability of the equipment but also the reliability of the personnel concerned.

CHAPTER 13 **PACKAGING AND TRANSPORTATION FOR RELIABILITY**

13.1 objectives

By the time the equipment satisfies the final factory tests, it should have an acceptable quality and reliability which has been achieved by the combined efforts of many people. Now we must get it from the factory to the customer with negligible damage. This is an important operation, and it is no use waiting until the equipment is ready for despatch and then instructing someone to 'find a case to fit it'. Packaging must be planned right from the design stage, with the following objectives in mind:

1 It must protect the equipment from damage during both transportation and storage.
2 It must, if required, be attractive to the customer. For consumer goods, an important part of sales promotion may be bound up with the attractiveness of the package.
3 Sometimes economies and other improvements in design and manufacture of the equipment are possible, if adequate protection is provided during transportation and storage.

It is convenient to consider the following separately, although in practice they are all bound up with one another.

1 Preservation.
2 Packaging.
3 Transportation and subsequent storage.

13.2. information required

In order to design suitable packaging we need to know at least the following:

1 The product's weight, shape and size.

2 The worst treatment it can stand without appreciable damage, for example:

a Shock and vibration.

b Maximum and minimum temperatures.

c Humidity.

d Resistance to chemical and physical influences.

3 The worst treatment the package is likely to receive, for instance:

a The method of transportation.

b The conditions of any storage.

4 When and how manufacture of the equipment is completed.

We need this latter information, because if possible we shall integrate packaging with the production line and final test. Too often we find in practice that, when manufacture is complete, the equipment is put on a truck and trundled, possibly through the rain, across the factory yard to the despatch department. There it is packed in a waterproofed case, which effectively seals inside all the water it has picked up on the way. No doubt some time later the reliability engineers will be puzzling as to why the equipment has failed after only a short time in service!

13.3. preservation

Before packaging proper, we must consider what preservation the equipment itself needs. Any dirt or damp left during manufacture should be removed. Trichlorethylene vapour is often used for this, although white spirit, naphtha and kerosene are alternatives depending upon what our product will stand. After cleaning, we must ensure that it is thoroughly dried and possible methods include hot air ovens at say 150°F or 65°C, infrared ovens or clean dry air blasts.

Once all the grease and moisture has been removed, our product is in an exposed condition and we must apply a protective layer as soon as possible. This might consist of any of the following:

1 Paint type preservatives, such as plasticised resins or lanolins

in a volatile solvent. The harder the resin sets, the better the protection, but of course the more difficult it is to remove when required.

2 Grease type preservatives, such as soap thickened mineral oils.
3 Oil type preservatives. These are usually easy to apply but are a bit inclined to drain off vertical surfaces.
4 Strippable coatings. These are often applied by a hot dip and give very good protection. They do however tend to be expensive.

Any of the above may need to be backed up by a barrier material to keep the preservative in place. Where the surface of the product is very uneven, we may find it very difficult to get the preservative on to all the internal surfaces and, even if we do succeed, our customers may not appreciate the job of getting it all out again. A possible alternative may be to seal the whole product in an envelope made of polyethylene sheets.

13.4. packaging

There is no exact point where preservation ends and packaging begins. Even quite large objects such as tanks may be stored in plastic envelopes without any crating. These have proved to be watertight, and even partially airtight. They do however provide no protection against damage received during transportation, or dropping and similar impacts received during mechanical handling.

Clearly we need to know what sort of treatment the product is liable to receive. Some general tendencies are that:

1 Small packages are more likely to be dropped than large ones.
2 While packages are often dropped a few inches, they are seldom dropped from a great height.

We may need some cushioning material, such as felt, wood wool or shredded paper if the load is not too heavy. Otherwise sponge rubber or sheet rubber may be better.

We shall probably require an outer container, designed to meet the conditions we listed in section 13.2. Throughout, however, we shall have to consider costs. Special containers which are not returnable tend to be very expensive and it may be worth our while to consider a range of returnable containers.

13.5. transportation

During packaging and subsequent despatch, the product must be mated up with its paperwork. Failure to do so correctly can result in a customer receiving a different model from that which he ordered. Worse still, he may attempt to use it with disastrous results. As far as the customer is concerned, he has received a piece of equipment with zero reliability and a complaint is inevitable.

There are broadly seven methods of transportation as follows:

1 Air.
2 Post.
3 Passenger train.
4 Goods train.
5 Carriers.
6 Sea.
7 Own transport fleet, provided it is fully utilized.

In deciding which to use, we shall no doubt select the cheapest, consistent with the required speed of delivery and the vulnerability of the goods to damage. This in part links back to packaging since, if the package gives sufficient protection, it may be possible to use a cheaper mode of transport. On the other hand, if the package is light enough, air transport may be attractive.

We shall discuss storage of equipment prior to use in section 15.3.

PART THREE **RELIABILITY AND THE CUSTOMER**

CHAPTER 14 THE PURCHASE OF EQUIPMENT

14.1. making purchases

We are all customers for domestic goods. Basically we decide what we want—this is the equivalent of our purchasing specification—and then we attempt to get it at the lowest possible price. If we cannot find exactly what we have in mind, or prices are higher than we expected, we may decide to go without.

The purchaser for industry, institutions or the armed forces is in a similar position, except that he is buying on behalf of an organization so that his own personal preferences are of no matter. Instead he must estimate the probable benefit to his organization and balance these against the cost of ownership.

14.2. the cost of ownership

A man buying a second hand car may well refuse an offer of the model he wants, at the price he has in mind, merely because he is not satisfied about its condition. He concludes that it will cost a fortune in repair bills. This applies equally to expensive capital equipment. The cost of keeping it maintained and in repair may well exceed the original purchase cost, during every year of its life. We must therefore consider the total cost of ownership in its widest sense and the following are some of the factors involved:

1 The purchase price of the equipment concerned, plus any ancillary equipment which may be required to operate it.
2 The cost of transporting the equipment to the site and installing it there. Included here may be the cost of foundations and of a building to house it.
3 Commissioning costs. The equipment must be adjusted and brought into service. Probably the purchase price will include the services of an engineer from the supplying com-

pany, but it is a matter of experience that large non-standard systems seldom do work initially exactly as the designer intended. We must therefore be clear from the start who is going to pay for any modifications which have to be made to correct teething troubles, for their cost may be very considerable.

4 Maintenance and repair costs. These may well be the really expensive items, and they go on for year after year.
5 The cost of training operators.

14.3. performance and reliability

We must ensure that the required performance and reliability are fully and clearly agreed with the manufacturers before the order is placed. We shall check their reputation and try to judge the extent to which their designers are reliability minded, and act upon the points set out in Chapters 6 to 11. We shall want to satisfy ourselves that the factory quality control system will ensure good conformance of the product with drawings and specifications. We may well ask to visit a similar installation made by the same company, but we must remember that we are sure to be referred to a satisfied customer, even though there may be plenty of dissatisfied ones as well.

Probably the reliability will be quoted as a failure rate or MTBF but, whichever method is used, the figure given will only be meaningful if a failure is clearly defined. This problem arises particularly with degraded failures. Thus if the system does not work as well as we expect, we may find that the manufacturing company say that in their opinion it is quite satisfactory. We now have an impasse, in which we as the customer will nearly always have to foot the bill if we want the performance of our system improved.

14.4. maintainability of the new system

Along with reliability, we must enquire into the maintainability of our new system.

1 SPEED OF REPAIRS

a We must satisfy ourselves that the proposed system has been designed as discussed in Chapter 11, so that maintenance and repairs can be carried out easily.

b We must investigate the availability of spares. It may be very costly to have the whole system out of service, merely because a vital spare cannot be obtained. We must consider the cost of holding a running stock of essential spares.

c The availability of maintenance personnel with the necessary skill is important. If we already have or can provide suitable men on our own staff, then they will always be available when required. Complex systems, however, are often such that repairs, other than simple replacements, require the services of a skilled man from the manufacturing company. This immediately takes the control out of the customer's hands. If may well be that the manufacturing company will always give us very good service, but we shall nevertheless be dependent upon them. There have been cases where the manufacturing company has later gone out of business, leaving all their former customers to fend for themselves.

2 PERFORMANCE AFTER REPAIRS

Some delicate or complex pieces of equipment do not appreciate the disturbance caused by a major repair and never really work as well afterwards. In such cases, it is particularly important to purchase a reliable system in the first place.

14.5. losses caused by the non-availability of the system

Reliability and maintainability combine together as we saw in section 4.3 to give availability. As customers we shall be vitally interested because there are almost certain to be additional costs and inconveniences incurred whenever the system is not available for use.

1 PROFIT MAKING ORGANIZATIONS

Where our objective is to make a profit, then we can in principle work out the cost of not having the system available to us. Thus in the industrial case, losses may include:

a Cost of scrap or rework made at the time the system failed. To this we must add the cost of any materials which had to be scrapped because they could not be processed in time. Thus in a bakery, it might include the cost of dough which could not be baked due to an oven breakdown.

b Paying operators made idle by the breakdown. This may include not only the operators on the system itself but those on preceding or following operations who were also unable to work.

c The cost and nuisance value of split batches, extra setting time, etc.

d Loss of profits which the system would have made had it been working, plus the cost of unrecovered overheads while it was idle.

e Damage to good customer relations caused by late deliveries. This is almost impossible to cost, but it can be very expensive.

2 ORGANIZATIONS WHICH PROVIDE A SERVICE TO SOCIETY

Not all service organizations are non-profit making. A bus company can calculate the cost of a bus breakdown and the electricity boards the cost of a power failure. Where however organizations are financed by society to provide some service, the cost of unreliability can be very difficult to calculate. Thus if a fire engine breaks down on its way to a fire, we may be able to work out the cost of the building which was needlessly destroyed. But if in addition a life was lost which could have been saved, who is to put any meaningful cost on that?

The armed forces have a similar problem. If a vital piece of defence equipment fails just as an enemy approaches, it is unlikely that the loss can be expressed in terms of money. This might be possible, however, if the failure merely delayed the training of a group of men and so wasted their time.

14.6. return on a capital investment

In sections 14.2 to 14.5 we have surveyed the many outgoing payments which the purchase of a new system may involve. Collectively they are referred to as the cost of ownership. Against these we must set the benefits of ownership and in principle unless the benefits exceed the cost by an acceptable margin then the purchase should not be made.

1 PROFIT MAKING ORGANIZATIONS

In profit making organizations, the benefits usually come in the form of increased profits. Thus the money paid out, called by accountants the cash outflow, can be balanced against the money received or cash inflow as a result of possessing that particular system.

2 ORGANIZATIONS WHICH PROVIDE A SERVICE TO SOCIETY

It is very difficult to find a suitable yardstick for assessing the value of a non-profit making service to society, but nevertheless some logical attempt should be made. Most such organizations are restricted by a budget which sets an upper limit to what they can spend each year, and they then have to choose between a number of rival and maybe very worthy claims on their funds.

One approach is to compare the situation if the equipment concerned is bought with the situation if it is not. Thus a university or college can consider not only the benefits to students or research which will accrue from the purchase of a particular piece of equipment, but also exactly how much each will suffer if that purchase is not made. One should consider quite realistically how far the purchase will improve the service being offered, and how far it is really a status symbol to enhance the reputation of the organization concerned.

The armed forces have if anything an even more difficult task. Consider the difficulties of putting a value on a piece of defence equipment intended as a deterrent. Suppose the equipment is bought but never sees active service, because no enemy attacks during its useful life. Does this mean that the equipment has proved its worth in deterring the enemy or was it a complete waste of money because the enemy never intended to attack anyway?

14.7. the decision whether to purchase

In principle, once the cost of ownership has been determined and the profits which will accrue have been estimated, it is a straightforward if sometimes tedious task to decide whether or not to purchase a particular system. Industry uses three methods to make this decision.

1 THE RETURN ON CAPITAL

In this method we calculate the total cost of owning the system, and the total cash inflow which will be received from it.

Thus suppose the cost of ownership	= £40,000 over 10 years
The total cash inflow	= £60,000 over 10 years
Total profit	= £20,000
Average profit per year	= £2000
Percentage return on outlay	$= \frac{2000}{40,000} \times 100\%$
	= 5%

The objection to this method is that it takes no account of when money has to be paid out or of when it is received. Probably most of the £40,000 was paid out at the start but the £60,000 might have been received right at the end of the 10 years, leaving a lot of capital tied up for a very long time.

2 PAYBACK PERIOD

For this we estimate how long it would take to recover our outlay, the assumption being that after that we start to make a profit. The risk is that although a project may recover its outlay very quickly it may then come to an end and never make any profit.

3 DISCOUNTED CASH FLOW

This is the only one of the three methods which takes full account of the financial situation. It is a little more complicated than the others, and we shall not go into detail. Basically it is argued that we should consider not only how much but also when we pay out or receive money. The rules are :

a The later we have to pay out money the better.
b The earlier we receive money the better.

Thus suppose that in 2 years' time we have to pay £10,000 for a piece of equipment. We could invest a suitable sum of money now to pay that bill. If we can get 10 per cent compound interest, we need only invest £8,260 now, to receive £10,000 in 2 years' time. If of course we are to receive the money, the reverse applies. If we can have the money to invest now, in 2 years' time it will be worth £12,100.

A	B	C	D	E
YEAR	CASH OUTFLOW		CASH INFLOW	
	As paid	*Discounted to year 0 at 8%*	*As received*	*Discounted to year 0 at 8%*
0	£1940	£1940	0	0
1	£1600	£1482	£1000	£926
2			£1000	£857
3			£1000	£794
4	(−150)		£1000	£735
		£3422		£3312
Deduct £150 scrap value, discounted 4 years		110		
		£3312		£3312

The discounted cash flow method is illustrated above and the steps are briefly as follows.

a THE CASH OUTFLOW All the money we must pay out is set down against the year when we must pay it, as shown in columns A and B. The chief items are likely to be the cost of plant and equipment, buildings to house them and possibly working capital to start the project off. In year 4 we have entered −£150. This is the scrap value of the plant and equipment, etc. and it is negative because this sum will be received and not paid out. The government may offer incentives to encourage investment in certain types of equipment in specified areas. If any of these apply, we shall deduct them from the figures before they are entered in column B.

b THE CASH INFLOW In column D we enter the net cash inflow. For simplicity we have shown the same figure of £1000 for each of the 4 years. In practice they are unlikely to be nice round figures because they must take account of all of the following factors:

1 The profit which accrues from the project. To calculate this we must deduct all expenses, and this will include maintenance and repair costs as well as all other operating costs. From the profit we shall deduct the Corporation Tax which we shall have to pay on it.

2 We shall however add in the saving in Corporation Tax which accrues from the investment itself. Having spent £3540 on plant and equipment, we shall depreciate it each year. This really means that we shall charge our customers for the extent to which making their product wears out our machine. As these sums go to eventually replace the machine, they will not be counted as profits and hence there is a saving in the Corporation Tax which we have to pay.

c THE SOLUTION RATE OF INTEREST The sums of money in columns B and D are now discounted back to year 0. This means that we find their value in year 0 along the lines already discussed for our example of £1000 in 2 years' time. Our object is to find a suitable rate of interest such that, after discounting back to year 0, the total of column C equals the total of column E. This is called the *solution rate of interest*, and it represents the true earnings of the project. In the example the solution rate of interest is 8 per cent. We shall probably need one or two trials before we hit upon the right value, but for clarity only the correct value is shown.

4 THE ACCURACY OF DISCOUNTED CASH FLOW CALCULATIONS

Since discounted cash flow is obviously more complicated than the other methods, it is fair to ask just how accurate it is. As with so many management techniques the method itself is very accurate, and in practice everything depends upon the accuracy of the data which is used for the calculations.

By the end of year 4, all the figures will be known and our accountant will be able to tell us the actual solution rate of interest. By then however we are dealing with history and, if we find that the project has made an overall loss there is nothing we can do about it except learn for next time.

It is at the beginning of year 0 or even earlier that we want to know the profitability of the project, so that we can decide whether to go ahead or abandon it. Unfortunately at this stage the cash inflow figures particularly are probably little more than guesses. We cannot really know how well a new product will sell or whether its life will be curtailed by a rival one from a competitor. The cash outflow figures are probably reasonably accurate on purchase and installation costs, but it will be difficult to foresee the cost of teething troubles, possibly requiring extensive modifications, and leading to expensive delays in commissioning the new equipment.

The above is not of course an argument against the use of discounted cash flow, but rather a warning that an accurate method will not give an accurate result unless it is fed with accurate data. Readers who wish to know more about this technique may care to read *Appraisal of Investment Projects by Discounted Cash Flow*, by A. M. Alfred and J. B. Evans, published by Chapman & Hall.

CHAPTER 15 INSTALLATION, STORAGE, COMMISSIONING AND OPERATION FOR RELIABILITY

15.1. installation

Once we have decided to purchase a piece of equipment, we must plan for its installation. This can be quite complicated and involve :

1 The erection of a suitable building to house it.
2 Preparation of a base or foundations to receive it and the provision of services such as gas, electricity, water, drainage and compressed air.
3 Receipt of the equipment and if necessary its storage elsewhere until its permanent site is ready.
4 Installation.
5 Commissioning.

Items (1), (2) and (3) are usually the customer's responsibility, although they may be done to instructions and drawings supplied by the manufacturer. The latter usually supervises items (4) and (5).

Timing now becomes important. We cannot for example install the equipment until the building is ready and much therefore depends upon whether this is completed on time. If it is late the whole project will be upset. Temporary storage will have to be found for the equipment when it arrives, and installation and commissioning will have to be postponed. This however will probably upset the manufacturer's plans, and we may well find that their engineers cannot be made available at the revised later date because by then they are due elsewhere. Hence further delays follow and we do not have the use of our new equipment until long after we had planned. Of course it is not only the builders who can upset things by being late, but because the building comes so early any delay is likely to dislocate other parts of the plan.

It is usual to use critical path analysis to plan a large project of this sort. Basically we draw either a network or a bar chart which shows the earliest date on which each section of the work, called an activity, can start, bearing in mind that often one activity cannot begin until others have been completed. It will also show the latest date on which each activity can start if the whole project is to be completed on time. The first overall analysis may then be broken down further. Thus the building may have a critical path analysis of its own, showing commencing and finishing dates of such activities as clearing the site, foundations, sewers, walls, roof, etc.

As reliability engineers, we shall watch every stage of the installation for things which might cause failure later. Experience in the type of work involved will help here, and the following are merely examples:

1 If the equipment is to work alongside a chemical process or near plant emitting vibrations, etc. consider whether it needs any protection. A computer controlling the sequence of processing on, say, a plating line, will need to be near the plant concerned, but it must be protected from acidic and other fumes.

2 If the plant to be installed emits any effluents, we must make adequate provision for dealing with them. It is no longer permissible to tip any liquid effluent into a sewer, nor can we discharge all gases into the atmosphere. If the plant has to be shut down later, while suitable effluent treatment equipment is installed, it will be out of commission just as much as if it had broken down.

3 We must keep a close watch on the position of pipes and electric cables. For instance, we must:

a Avoid exposed positions where the weather may cause damage. A frozen water pipe may cause a complete plant to be shut down.

b Make sure that pipes and cables are not put into inaccessible places, so that later on repairs are difficult resulting in the plant being kept out of use longer than it need be.

c Electric cables may suffer from corners which are too sharp, positions where the wires get chaffed, wrong cleats, etc.

15.2. preparing for the operator

Early preparation must be made for the operator. If the system is of a new type as far as we are concerned, then those who will operate it must be trained. For simple items of equipment, an instruction period with the engineer who comes to commission it may be all that is necessary. For complicated systems, much more formal training will be required. Often the manufacturer will provide this, but basic training may be arranged with a local technical college. Even though he may not be directly responsible for it, the reliability engineer has a deep interest in seeing that this training is properly done because failures may be caused by untrained operators every bit as much as by an unreliable system.

Timing of this training is important. It should be comfortably completed before the system has to be operated. If however it is carried out too early, then those trained may well have forgotten a lot by the time the system is ready for use.

Safety is also vital. Although the manufacturer has some responsibilities for the safety of the equipment he provides, the real responsibility rests with the company operating it. Therefore we must consider whether any safety precautions are required and if so put them in hand in good time. It is most infuriating to discover that a newly installed system cannot be used because its safety guards are not ready.

We must also consider working conditions. Probably the designer of the system will already have given this some thought, along the lines discussed in sections 10.1 and 10.2. It is our job as user, however, to make sure that the operator is comfortable, and able to give of his best. The following are some of the factors to be considered.

1 WORKING TEMPERATURE In a factory the room temperature must reach 60°F within 1 hour, if the work is largely sedentary. In fact, however, 60°F is somewhat chilly, and at least 65°F is desirable.

2 LIGHTING No one can do a good job if he cannot see properly, so adequate lighting is a good investment.

3 NOISE AND OTHER DISTRACTIONS In general it is better to reduce the noise level emitted, possibly by soundproofing the offending machine, rather than to give the operators ear muffs. It is usually operators doing other jobs nearby who find a noisy machine irritating. The person who works the machine itself is seldom worried.

4 SEATS Where a seat would be useful it should be provided and made adjustable to a convenient height. It is no good catering for the average man or woman; they come in all shapes and sizes.

5 PROTECTION FROM THE SYSTEM ITSELF Depending upon its type, the system we are installing may emit heat, fumes, etc. and we must protect the operator in the ways discussed in section 10.2.

In general it pays to give careful consideration to operator comfort, since this encourages interest and cooperation and helps to avoid system failures due to the operators' inattention or inexperience.

15.3. storage

Storage of equipment may be necessary at a number of stages. Thus:

1 The manufacturer may store:
a Partly made products, which are awaiting the next operation.
b Completed products ready for despatch.

2 As the purchaser we may need:

a A temporary storage area, where we can keep the packages containing the system itself between the time of their arrival and installation.

b A permanent store for spare parts and replacement units, etc.

Perfect storage is practically impossible. However good our storage conditions, there are almost always slow changes going on in the equipment stored, and usually it is for the worse. Thus if a batch of parts which are all working perfectly are put in store, later when they are taken out we shall probably find that a few have failed. Indeed we can think of the parts as having two different failure rates:

1 A failure rate when operating under normal conditions.

2 A failure rate when idle in store.

Deterioration during storage may take the form of general loss of performance more than of sudden failure.

It is almost always better to store under correct conditions rather than to rely on tests afterwards to detect the failures. An interesting example of this occurs in the storage of ammunition. Thus suppose we have a large consignment of shells, which we hope will never be required for use against an enemy. Nevertheless, if the need should arise, we must be as certain as possible that they will fire as intended. Good storage conditions are the only real answer. We must find out what conditions of heat, humidity, etc. best suit the shells, and then provide it. Our control must be on the conditions of storage, which we can check say daily, and correct as necessary. We may also carry out visual and other non-destructive tests on the shells themselves. Of course the ultimate check on the reliability of a shell is to fire it, and observe whether it explodes correctly. But this should be used very sparingly. Firing one shell tells us very little. For a meaningful test, an appreciable number must be fired and, if we do this at intervals over a long period, we may find that our stock of shells has been significantly reduced.

When equipment arrives from the manufacturer, it is usually desirable to check that the right type has been sent and that it is in good condition and working order. We must take great care, however to make sure that we do not disturb any protective packing. Otherwise we may find that equipment which was perfectly satisfactory on receipt fails to work a little later when we require it for use.

15.4. commissioning a new system

Commissioning itself is usually done by the manufacturer's engineer, using a fairly precise routine, which has been developed for that type of equipment. Large systems are never started up in one grand 'switch-on'. Although the method varies, it is usual for each section or unit to be checked out and started up individually. The objects are :

1 To avoid damage should anything be wrong.
2 To pin point the location of anything which needs adjustment or correction.

A large aircraft will be extensively checked out on the ground, during which each part of its system will be made to function individually. It will then do various taxi runs on the runway, and only when everything has been shown to be working correctly will it be allowed to take off.

Once commissioning is complete as far as the manufacturer is concerned, we have what is probably our best opportunity to check the performance of the system against its guaranteed specification and, if it falls short, to get the manufacturer to adjust and modify it as necessary. Checking its reliability however is not nearly so easy, and although in Part Four of this book we shall discuss acceptance tests on parts and sub-assemblies, meaningful tests on complete systems are usually almost impossible. Probably we shall decide to accept the system on its performance, with its reliability still unchecked, and rely on the reputation of the manufacturer as an assurance that any shortcomings will be put right. Sometimes, however, it is possible to have a system 'on trial', for a few weeks, in which case it will only be accepted and paid for if it gives satisfaction.

If the system is of a new type, we shall want to check its performance in detail. For example, a new type of numerically controlled machine tool should have its accuracy checked for a variety of operations, different materials and other conditions of work.

Finally we must consider whether any further instructions are necessary to help the operators. They should not constantly have to refer to the instruction manual. It may help to add additional information to some of the controls. In particular, if any alarm signal operates, they must know at once what it means and what action they have to take.

15.5. operating the new system

If our new system is merely a duplicate of one we already have, we shall know its potential capabilities and checks to make final adjustments will be all that are required. Even so, we shall no doubt watch it closely for a week or two until we are satisfied with its performance and reliability.

Our commissioning checks will probably merge into and eventually become the quality control which will be used throughout its life. In most industrial cases, the new system will itself manufacture other products, and so its quality control will be broadly as discussed in Chaper 12. Indeed we have now come full circle, because what we produce must in turn satisfy the quality and reliability requirements of our own customers.

If we are concerned with a service organization, we must still ensure adequate operational quality control, as outlined in section 12.10.

15.6. new generation systems

If we decide to buy a system of a new type, either designed specially for us or the first of a manufacturer's new production, then we cannot expect everything to go as smoothly as it should for a well established system. We are deliberately aiming for improved performance and we shall probably have to pay for it, at least initially, by accepting a lower standard of reliability.

We must therefore arrange for very close control until:

1 We know precisely what its performance is and are satisfied with it.
2 Its reliability is adequate to be left to routine control.

In section 2.2 we noted that parts and systems tend to exhibit a higher failure rate during the early failure period at the start of their lives. This may be due to manufacturing defects which have not been detected in the factory or to design weaknesses. A new design is particularly liable to suffer from such failures, and this may last for several hundred operating hours. The actual pattern of failure will vary with the system, but it is often found that the Weibull distribution can be made to fit it provided suitable constants are chosen, as will be explained in section 20.11.

Once the new system has been established, we shall want to estimate how useful it is for our particular work and two quantities may then be useful.

1 TIME UTILIZATION This is the proportion of the total time during which the system is available for use, that it is actually being usefully used. A factory will normally hope that this will be very high, but in some industries the time utilization is bound to be low. Thus on a farm a combine harvester can only be used for 2 to 3 months per year.

2 TECHNICAL UTILIZATION This is the extent to which a system's technical capabilities are being used. It is easy to purchase an advanced item of equipment and then discover that there is very little work available which really needs its special capabilities. Hence it spends most of its time doing work which a much less sophisticated, and therefore cheaper, equipment could do equally well.

CHAPTER 16 MAINTENANCE FROM THE USER'S POINT OF VIEW

16.1. the mean time to repair

When we buy an item of equipment, we in effect take over from where the designer and the factory left off, our object being to make that equipment available for full use whenever required, while keeping maintenance costs to a minimum.

Suppose we have a considerable number of equipments of the same type and that each time a failure occurs we note how long it takes to put it right. When we have enough information, we could average all these times together and estimate the mean time to repair (MTTR), which we mentioned in section 4.3.

Observed mean time to repair = MTTR

$$= \phi = \frac{\text{Total number of maintenance action hours}}{\text{Number of maintenance actions}} \qquad (16.1a)$$

This is in everyday language:

$$\text{MTTR} = \frac{\text{Total number of hours spent on repairs}}{\text{Total number of repair jobs}}$$

Mathematically the MTTR is similar to theMTBF of section 2.9, although it has an opposite significance in that we wish:

1 The mean time between failures to be as *long* as possible.
2 The mean time to repair to be as *short* as possible.

Similarly the reciprocal of the MTTR corresponds to the failure rate, so we can write:

$$\frac{1}{\text{MTTR}} = \mu = \text{Maintenance action rate}$$

= Average number of equipments which can be restored to normal working per hour, assuming that there is no waiting between repair jobs.

In the same way that $$\lambda = \frac{1}{\theta} \qquad (2.9a)$$

So $$\mu = \frac{1}{\phi} \qquad (16.1b)$$

Whereas we want the failure rate to be *low*, we want the maintenance action rate to be *high*.

16.2. distribution of times to repair

Since the actual time taken for each maintenance action may vary considerably, suppose we plot a frequency distribution curve as shown in Fig. 16.2(a).

We should not expect a symmetrical frequency distribution curve like that in Fig. 3.5(a), because this would imply that maintenance action times were just as likely to take a little longer than the MTTR, as to take a little shorter. In Fig. 16.2(a), the most likely duration of a maintenance action is given by the peak of the curve, called the mode. Some jobs will take less than this, but there is a limit because they cannot take zero or negative

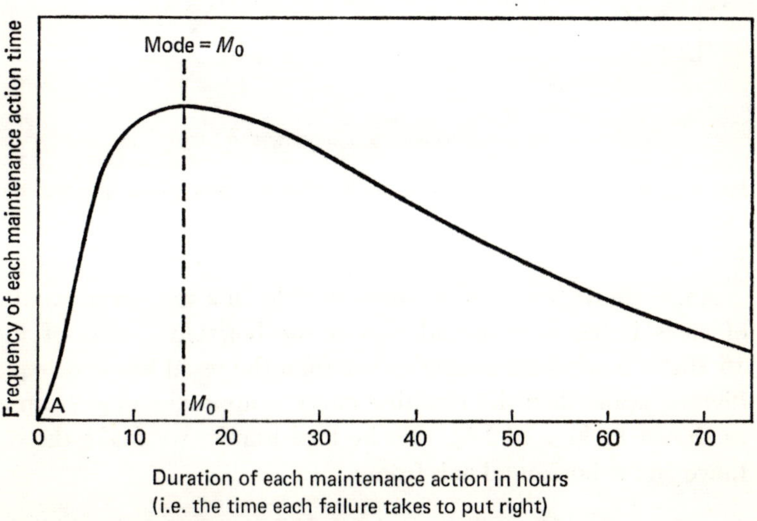

FIG 16.2(a). Frequency distribution curve showing the frequency of each maintenance action time

times. Hence all the values below the mode must be confined to the section of the curve marked AM_0. Above M_0 however, there is no such limitation. Now and again, a breakdown may take a very long time to put right, and so we usually get a skewed curve as shown. Because of this, the MTTR or average is unlikely to coincide with the mode and will normally be to the right of it.

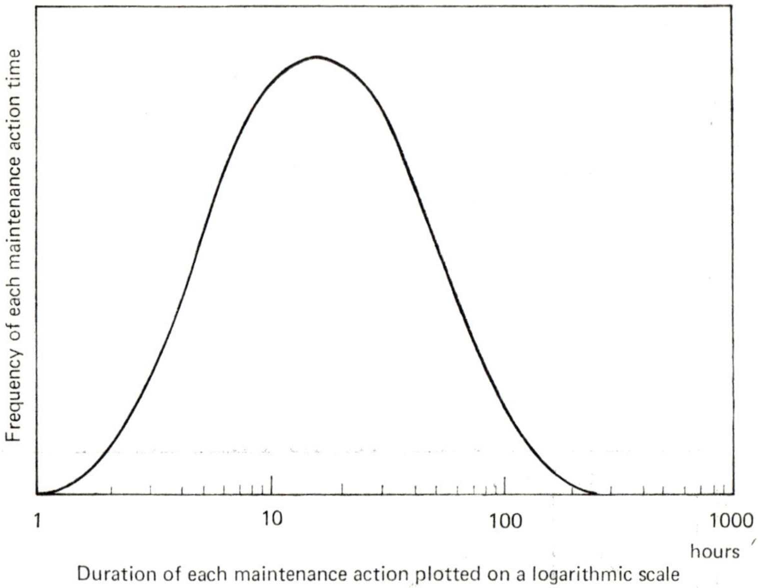

FIG 16.2(b). Frequency of each maintenance action time, plotted to a logarithmic scale

Although there can be no absolute rule, in a very large number of cases it has been found that if the horizontal scale of Fig. 16.2(a) is made logarithmic rather than the usual linear or equal division scale, then the resulting curve is approximately Normal, as shown in Fig. 16.2(b). (To be logarithmic, we make the distance on the horizontal axis from :

0 to 10 = distance 10 to 100
= distance 100 to 1000 and so on.)

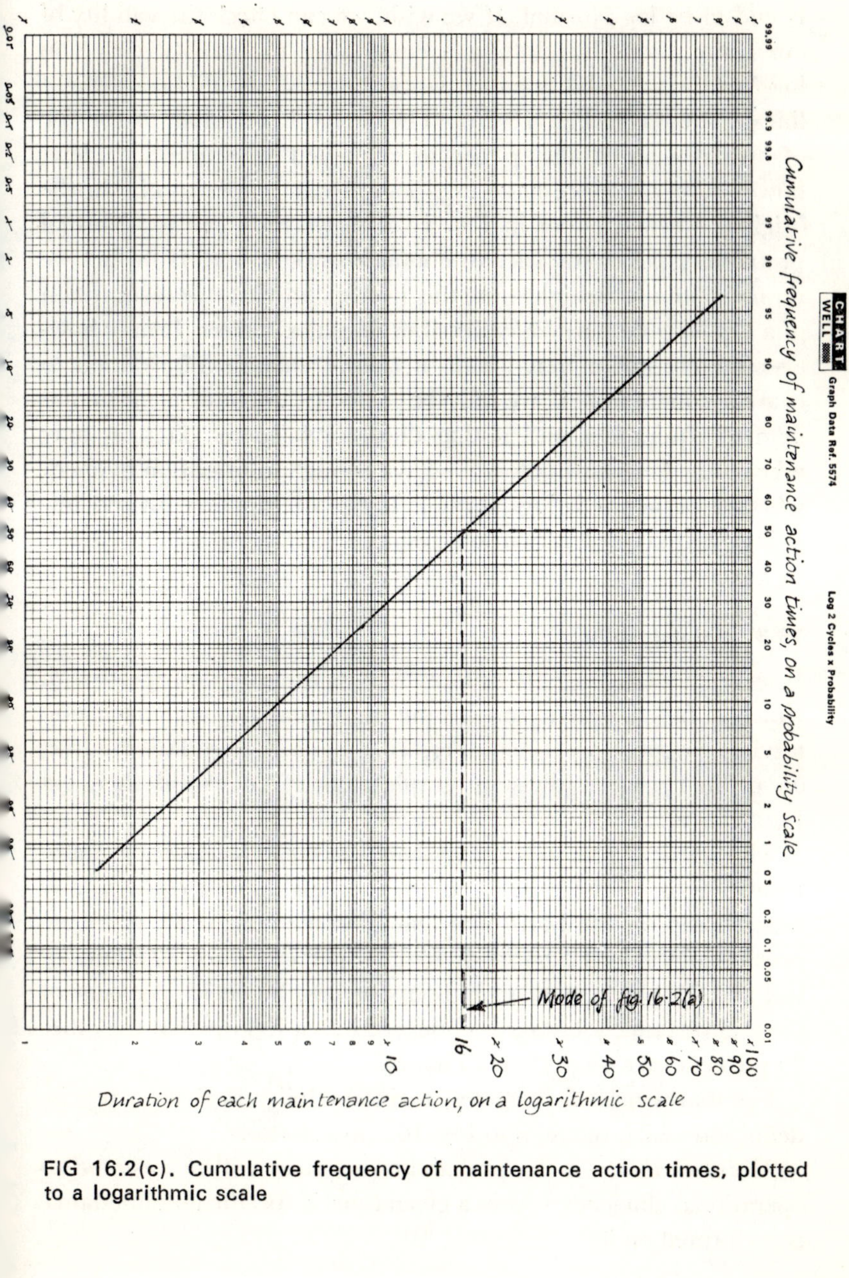

FIG 16.2(c). Cumulative frequency of maintenance action times, plotted to a logarithmic scale

Where this relationship holds, the frequency distribution curve is said to be log-Normal. If we wish, we can check the validity of our assumptions by plotting a cumulative probability curve on log-Normal probability paper, when the result should be a straight line as shown in Fig. 16.2(c). The method is similar to the use of ordinary probability paper, as explained in section 3.9. Here, however, because we have used logarithmic paper to obtain the original Normal curve in Fig. 16.2(b), we must use log-Normal probability paper in order to obtain a straight line. If in fact we do not obtain a straight line, this may mean that our assumption of a log-Normal frequency distribution is not correct. It may however merely mean that we do not have enough information to draw a conclusion. If we only have one example of a particular item of equipment and if that does not fail very often, it may take a very long time before we have enough information to check whether or not it is log-Normal. If we attempt it with too few points, we may well get curious results.

16.3. maintainability

In everyday language, maintainability is the ease and speed with which a failed system can be put right and brought back into service. Suppose we replot Fig. 16.2(a) as a cumulative curve, on ordinary equal division graph paper. We would then get Fig. 16.3(a).

The vertical axis now represents the total fraction of maintenance jobs, whose time to complete is less than or equal to any time chosen on the horizontal axis. For example, 0·25 or 25 per cent of jobs take 18 hours or less, 50 per cent take 37 hours or less and 75 per cent take 75 hours or less. Since 75 per cent of the maintenance jobs take 75 hours or less, it follows that the other 25 per cent take more than 75 hours.

For mathematical purposes, maintainability is given a precise definition which relates it to Fig. 16.3(a) as follows.

Maintainability is the probability a system will be restored to operational efficiency within a given time, if specified maintenance is performed on it.

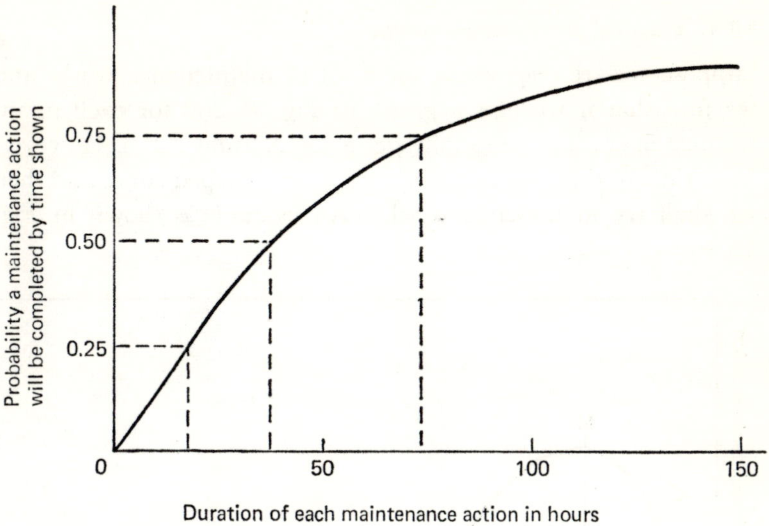

FIG 16.3(a). Cumulative curve showing the probability that a maintenance action will be completed within a given time

Maintainability is thus similar to reliability,* since :

> *Reliability* is the probability a working system will not fail within a given time.
> *Maintainability* is the probability a failed system will work again within a given time.

For reliability we had to state the conditions of use and now we have to specify the maintenance which is to be performed. Clearly if we decide to put two men on to a job instead of one, we should expect to increase the probability that the job will be completed within a given time.

The 'given time' in the above definition, is called the *maintenance time constraint*, and the greater this is set, the greater the chance that a given maintenance job will be completed within it. (This is merely stating that the longer we allow to do a job, the greater the chance that it will be done within that time.)

* Strictly speaking it is unreliability and maintainability which are mathematically similar.

16.4. practical maintenance curves

Suppose we are responsible for a lot of maintenance work, and we find that if we plot a graph as Fig. 16.2(a) for each maintenance operation we usually get a log-Normal curve. However, from time to time, we may get a peculiarly shaped curve and then we shall try to understand why. An example is shown in Fig. 16.4(a).

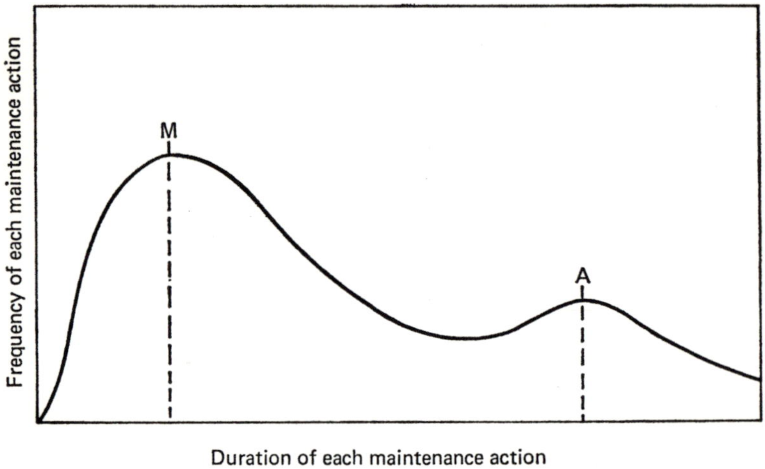

FIG 16.4(a). Frequency of maintenance action curve with two peaks

In Fig. 16.4(a) the curve is basically log-Normal, but there is a rise at A where we should expect it to continue to fall. This means that a proportion of the jobs are taking significantly longer than expected. Thus:

1 We may find that the times around A are mostly associated with a particular aspect of the maintenance work. Perhaps some part of a mechanism fails from time to time. Usually it is a straightforward replacement, but occasionally the part concerned disintegrates and falls into the rest of the mechanism and then the whole thing has to be stripped down and cleaned.

2 The rise at A may be associated with particular maintenance

personnel, rather than with the type of maintenance and this might indicate:

a Inadequate skill, resulting in an unexpectedly long time to complete work. This is not very likely, because we should expect the times taken by unskilled people to vary considerably and not to bunch at a particular point A.

b Delays due to non-availability of tools. Again, if the time the men had to wait for the necessary tools was variable, we should not expect their overall times to cluster about A.

c Wrong time booking. Maintenance staff may be booking on this job, time which is either wasted or should correctly be booked elsewhere. If the usual time for a job is a bit less than the length of a shift, we may find that the duration of the shift is being booked instead of the actual time. This would certainly account for times bunching at point A.

d There may be overmanning on some jobs. This is particularly liable to occur on large maintenance jobs.

3 Sometimes a second rise is so dominant that it amounts to a curve with two main peaks. This probably means that there are two distinct groups of maintenance work represented in the curve, each of which has a log-Normal distribution curve which is quite distinct from the other.

16.5. analysis of maintenance time

In our definition of maintainability, we referred to *specified maintenance* being performed. This includes the whole time for which the system is out of commission, and a typical sequence of events might go like this:

1 A failure is reported to the maintenance staff.

2 A maintenance technician collects together the tools and spares he thinks may be needed and makes his way to the failed system.

3 He verifies the precise way in which it fails to function. (In a few cases, he may find that there is really nothing wrong with the system, but that the operator is inexperienced.)

4 He locates the fault.
5 He obtains any spares, replacements, etc. which he needs. If these are not immediately available, there may be some delay while they are ordered and delivered.
6 He carries out repairs, and makes any adjustments which are necessary. This might involve:
a Replacement or repair on site.
b Substitution of a good unit for a faulty one, the latter being taken to the maintenance depot for repair.
7 He tests to make sure that the system now functions satisfactorily.

We shall naturally be anxious to get every failed system back into service as quickly as possible, yet a glance at the above shows that time can be eaten up by things such as ordering spares which are not part of the repair work itself. Indeed a second definition besides maintainability is sometimes used, and this is called repairability.

Repairability is the probability a system will be restored to operational efficiency within a given *active repair time*, if specified maintenance is performed on it.

It is often revealing to divide the total repair time into the following:

1 ACTIVE REPAIR TIME This is the time during which one or more technicians are actually working on repairs, adjustments, etc.

2 LOGISTIC TIME Logistics is a word which originated in the United States army, and it then embraced all the provisions, supplies and support required for the fighting men, even including the removal of the wounded to hospital. In reliability it has a narrower meaning, and logistic time means the time taken to obtain all the materials and spares, etc. which are required.

3 ADMINISTRATION TIME Under administration we include all down time which is not specifically allocated to either active

repair or logistics. It includes time lost because the maintenance staff are not available. There will be meal breaks, and some system failures are sure to occur very near to the end of a shift, possibly at the end of the working week as well. In such cases, the system may not be worked upon until the following Monday morning. Everything will depend upon the urgency. If the system is a vital piece of defence equipment, then sufficient maintenance technicians will have to be found so that they can work in relays and keep the repair work going for 24 hours per day.

Other administrative delays may be caused by having to wait for a maintenance technician to complete a previous job or by the time taken to raise paperwork.

16.6. unscheduled maintenance

In principle, maintenance can take two forms :

1 UNSCHEDULED MAINTENANCE Whenever a system fails to work as intended, then repairs and adjustments must be made to put it right. Since we can never foresee exactly when such a failure will occur we cannot plan for it, and so this is called unscheduled maintenance.

2 SCHEDULED OR PREVENTIVE MAINTENANCE It often pays to carry out regular checks on systems, and to replace parts which are approaching wear out. We shall discuss this in section 16.7.

As reliability engineers, we shall want to know why a system which is working correctly one moment should fail shortly afterwards. Often it is the failure of one part or unit which has caused the whole to fail, but sometimes the trouble centres round a whole group of parts each of which has worn or drifted away from its nominal value. Each one on its own would probably be all right, but several together add up to overall failure.

Our first clue as to their source, comes from the bath tub curve in Fig. 2.1(a).

1 EARLY FAILURES

If the system is a new installation, we may be dealing with early failures, in which case as the weak parts are replaced by normal strength parts, we shall expect the trouble to die away.

2 CONSTANT FAILURE PERIOD

During the constant failure period we may experience random failures. In Fig. 6.10(b) we noted that these would occur whenever there was an overlap between the frequency distribution curve for part strength and that for use stress. While this is true in theory, there are probably many other sources of random failures arising from variations in the conditions of use. Thus:

a There may be a large voltage surge at the moment of switch-on or switch-off, which sometimes causes a part to fail.
b Two controls which should be used in sequence may be sometimes operated together or in the wrong sequence. The operator then tries to correct rapidly and this results in some unusual and unforeseen switching sequence.
c A mechanical system may be accelerated or decelerated sharply, possibly combined with a large or an unbalanced load.

It by no means follows that some operator inefficiency will be involved although, if it is, he will probably not admit it thereby adding to the mystery as to why failure occurred. All of the above may be combined with drifts in the strengths of the parts. We can think of this latter as meaning that the bottom of the bath tub curve in Fig. 2.1(a) is not quite horizontal, but rises gradually as time passes. Alternatively we can regard it as due to a reduction in mean part strength, resulting from a drift to the left of curve $B_1M_BB_2$ in Fig. 6.10(b). Both concepts are fundamentally the same.

The reliability engineer must try to ensure that each item of equipment is operated correctly, and within its rating. Indeed for additional safety, we can try to avoid working too near to full rating. This is equivalent to the designer derating parts, as explained in section 6.9.

3 WEAR OUT FAILURE PERIOD

Where the wear out failure period is well defined as in Fig. 2.1 (a) then the majority of failures may originate from it. Suppose we are responsible for maintenance in a building with a large number of tungsten filament lamps.
Consider the case where we start off at one moment in time with a completely new set of lamps. If we neglect early and random failures, we shall have a failure pattern like that shown in Fig. 16.6(a). Wear out failures will begin at approximately time $(M-3\sigma)$ and be complete by about time $(M+3\sigma)$, as we discussed in section 3.6, where:

M = Mean time to wear out.
σ = Standard deviation about that mean time.

Now suppose that each time a lamp wears out we replace it with a new one. This means that our second lot of lamps do not all start life together, but instead their starting times are given by the distribution curve about M. Thus when this second generation of lamps begins to wear out, they will fail at various times because:

a Some lamps have a longer life than others.
b They did not all start working at the same time.

The distribution curve at $2M$ is therefore lower, and spreads more than that at M. It will not however spread twice as far, because of an averaging out effect. Thus in the first generation of lamps, there were a few with a very long life, which failed at $(M+3\sigma)$. When they were replaced, however, it would be very unlikely that the new lamps would just happen to be also ones with a very long life. It is much more likely that they would be nearer to average, about half actually having a below average life. Hence the new standard deviation is not 2σ but $\sqrt{2}\sigma$.

If we repeat the replacement process several times, then the frequency distribution curves become progressively lower, and spread further and soon begin to overlap. This means that before one generation of lamps have all failed, some of those that have already been replaced need replacing again.

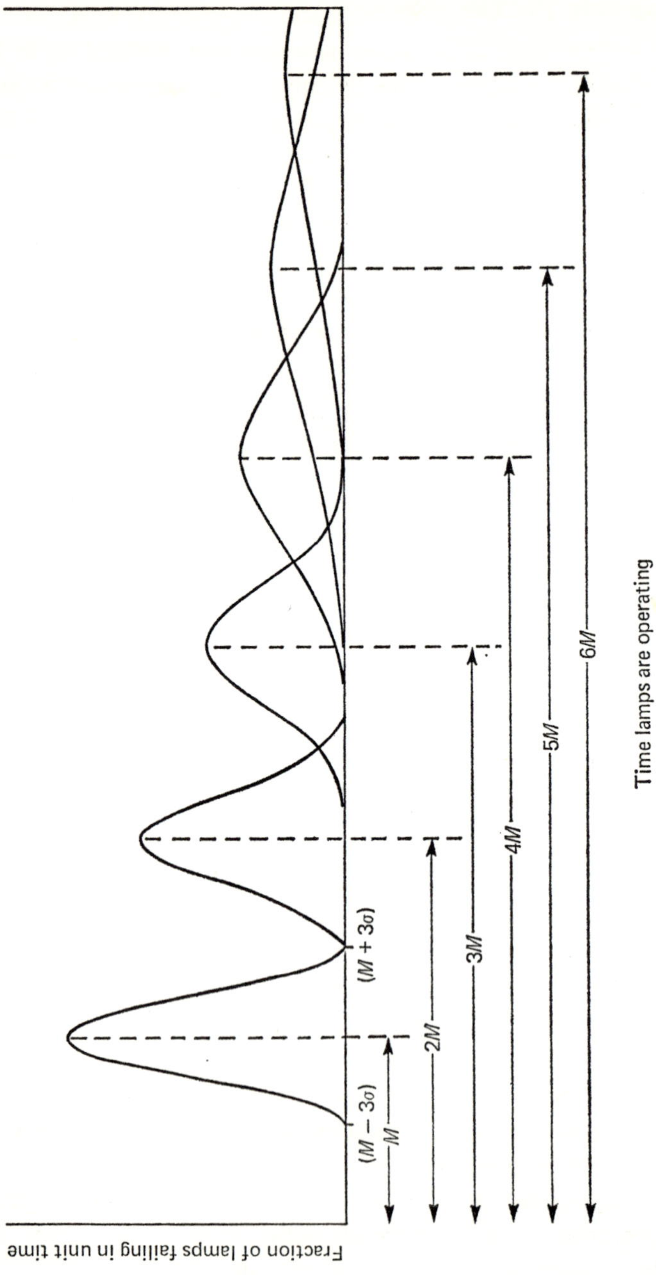

FIG 16.6(a). The effect of a policy of replacing items only when they fail

Eventually the failure rate becomes constant, and the Normal frequency distribution pattern is completely lost. Very roughly this happens in a time equal to $M^2/3\sigma$. However other factors will also tend towards a constant failure rate. Early failures and constant or random failures, from the first two periods of the bath tub curve, will occur in each generation of lamps, and as these will then have to be replaced at once some new lamps may start their lives at any time. Furthermore the extent to which each lamp is used may vary considerably, so that those which are used more often reach wear out long before those which are seldom used. Hence a constant failure rate may be established earlier than the above formula would suggest.

To the maintenance engineer, a constant failure rate may be very inconvenient. Suppose that instead of being in a building, our lamps were used to light the streets of a large town. It would probably be very expensive to replace lamps whenever they happened to fail. Our maintenance staff would spend long periods travelling across the town to a single failed lamp and then a relatively short time actually replacing it. Having installed a new set of lamps, it might well be cheaper to replace them all, whether

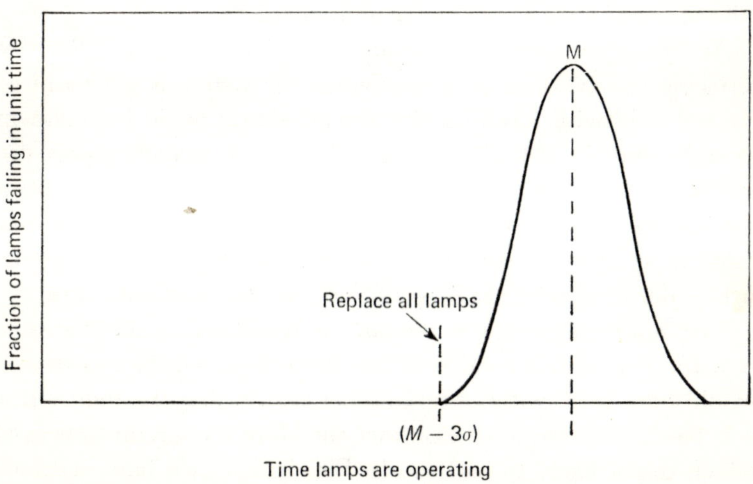

FIG 16.6(b). Replacement of all lamps together, to avoid having to deal with a constant failure rate

they had failed or not, at intervals of approximately $(M-3\sigma)$. Our object would be to avoid unscheduled replacements, and thereby to prevent the failure rate from becoming constant. This is illustrated in Fig. 16.6(b).

Whether this is a practical proposition depends in part upon how well defined the wear out curve is. If a large proportion of replacements are made necessary by early or random failures, then there is little to be gained.

16.7. scheduled or preventive maintenance

The above example brings us naturally to consider how far it would be useful and economic to plan our maintenance in advance. This is called *scheduled maintenance*, *preventive maintenance*, *planned maintenance* or sometimes *routine maintenance*. The basic idea is that at regular planned intervals we inspect the system concerned, replace any part which appears to be getting towards the end of its life and generally readjust so that it is restored to the best performance of which it is capable. A common example of scheduled maintenance is the regular servicing every thousand miles or so, which we give to our cars in order to avoid a breakdown at some inconvenient moment.

At first sight, scheduled maintenance is attractive. Routine servicing can be done at a time when the system is not required for use, and with any luck this should almost avoid breakdowns as such, and the risk of additional damage if perhaps a part disintegrates and falls into the moving or live parts of the rest of the system. In fact, however, the argument is not nearly so one sided as it appears, and we must consider all aspects carefully before deciding whether it would help in any particular case.

The main argument in favour of scheduled maintenance is that the risk of failure after it has been done will be less than it would have been without it. In other words, that the failure rate will be thereby reduced. Consider therefore the circumstances in which this is likely to be true. In Fig. 16.7(a), we have redrawn the bath tub curve from Fig. 2.1(a). Suppose that in the course of scheduled maintenance, we decide to replace a particular part even though it has not yet failed. The effect will depend upon just when we do it, for remember that, whenever we put in a new part, the failure rate curve will start again from time 0.

1 If we were foolish enough to replace a part at point A, the failure rate curve would immediately return to the beginning and follow line AUV before rejoining the original curve.

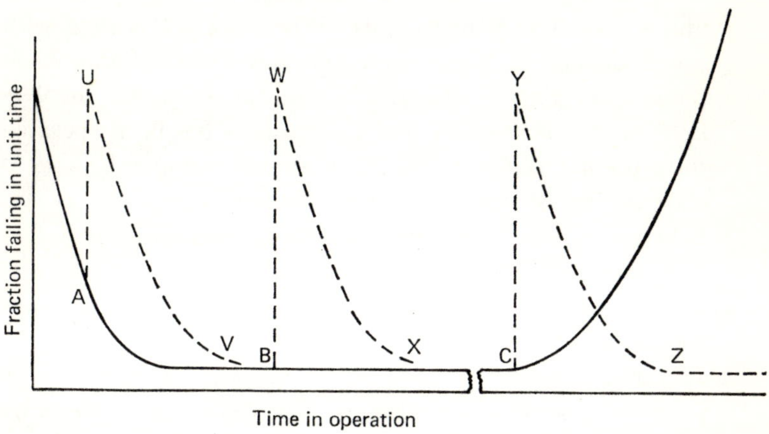

FIG 16.7(a). The effect on failure rate when parts are replaced at various times

2 At B the effect is worse. Parts which have reached the low failure rate of the middle period are discarded in favour of new parts, which will repeat the early failure period as BWX.

3 Only in the region of C is the substitution of new parts worthwhile, because after that the failure rate rises sharply as the old parts wear out, whereas the new parts will repeat the bath tub curve all over again and so return along CYZ to the low failure rate of the middle period. We could argue that C is a little early to replace, but we must remember that the failure rate of the old parts is rising rapidly and if it safer to replace a little early rather than a little late.

16.8. practical considerations concerning scheduled maintenance

We can now set out the main considerations which will decide whether we do scheduled maintenance or not on a particular item of equipment:

1 The failure rate after replacement must be lower than it was before.
2 It must be possible to predict that failure is likely to occur unless replacement takes place. Where we are dealing with a well defined bath tub curve, we may know roughly when point C of Fig. 16.7(a) will occur, but even this involves knowing for how many hours the system has been operated since the last replacement. Alternatively, we may be able to observe signs of wear in the parts themselves.
3 Scheduled maintenance may be useful to readjust the system to optimum performance.
4 The convenience of scheduled maintenance may far outweigh the inconvenience of breakdowns.
5 In many systems the parts adjust to each other, and settle down to a harmonious existence. The disturbance caused by scheduled maintenance can upset this.

Thus, in one case, analysis of army equipments revealed that those given scheduled maintenance had ten times as many faults as those without it. In another case, a group of electronic equipments were given scheduled maintenance every 1000 hours of service, until it was discovered that 30 per cent of all failures occurred within 24 hours of such maintenance.

To decide our policy in an actual case requires a balance between letting well alone and replacing that which is about to fail. Experience obtained from past maintenance on the type of system concerned is necessary, but a general guide can be deduced from the information in sections 2.5 to 2.7.

Many electronic parts hardly seem to wear out at all and the constant failure period of the bath tub curve goes on almost indefinitely. Where this is the case replacement is pointless. The inevitable disturbance to the wiring and circuitry in general will do more harm than good. Other parts such as transformers, inductors, resistors and capacitors tend to have very long lives provided they are not disturbed. Often they are best left alone and, if replacement is necessary, it may be better to replace a whole element or unit of the circuit rather than one part.

Mechanical equipments which suffer wear which can be observed before they fail are perhaps the most attractive subjects for scheduled maintenance, but even here caution is necessary. Parts which are externally corroded may fall to pieces if anyone tries to disturb them. Each case must however be judged on its merits.

16.9. maintenance records

Since an efficient and economic maintenance policy must be based on the way the equipments concerned do or do not fail, we must have some method by which we record and benefit from past experience. Form filling on a large scale however is expensive, and the information collected is of no value unless it is analysed and action taken on it. Maintenance technicians tend naturally to be practical men, who are more interested in getting on with the next breakdown than in filling in a form about the one they have just done, and this leads to forms being filled in from memory, perhaps once a week. Inaccurate information results, and this can be worse than no information at all.

Before adding to the paperwork, it is therefore worthwhile to consider how much information is either already available or could easily be made available. The following are examples.

1 An analysis of requisitions to stores for replacement parts, gives an immediate indication of:

a Which types of part are failing.

b In what quantity.

c Whether there are any seasonal or other patterns. It may be useful to add other boxes to the requisitions to show:

d The machine or equipment number on which the part will be used.

e The department where the machine or equipment is housed.

2 Maintenance technicians frequently have to keep time sheets, which are used to calculate overtime and other payments. On these they can, and probably already do, show how long they have spent on each job. This gives some indication of the active repair time.
3 In industry, the time a production operator spends without work because his machine has broken down is usually recorded.

Neither (2) nor (3) above indicates how long a machine is out of commission, item (2) because it does not include logistic or administrative time, and item (3) because the production foreman will do his best to find the operator another job to do while his machine is being repaired.

The true availability of an item of equipment can seldom be found from existing records. When a new type of equipment has just been installed its availability may be of considerable interest and it may be worthwhile putting a timer and recorder on it.

The accuracy of some of the foregoing records may be suspect. For example any new parts smuggled out for private use or time wasted will no doubt be booked to some equipment or other. Nevertheless there is a built in incentive to keep the records going. Thus new parts cannot usually be obtained without a requisition giving the information required, wages cannot be paid without the time sheet and so on.

When we have surveyed all existing documentation, and added any questions to them where necessary, then we can consider additional records for any vital but missing information. There is no guarantee that these specially designed forms will be filled in any more accurately than the existing ones. Indeed as there will in general be no particular incentive, the accuracy could be worse. Much depends upon their design. They should be :

1 As brief as possible, consistent with collecting the required information.
2 Quite clear about what is required. If the operator thinks, 'What on earth does this mean?' then it is a bad form.
3 As easy as possible to fill in. Ticks and crosses are easier

than having to write a description, although they have the disadvantage that maintenance men may get to know the anticipated answer and give this rather than the true one, believing that the latter will lead to queries.

4 As easy as possible to analyse. If the analysis is going to be a big job consider punched cards, so that they can be analysed by computer.

Throughout the design of the form, we must consider the use which will be made of the information and anything of little use should not be collected. The sorts of questions we shall want to be able to answer from the analysed results are:

1 Precisely what scheduled maintenance should be done on each item of equipment?
2 How often should it be done?
3 When is the equipment not required for operational use?
4 What are the chief weaknesses of each item of equipment and what can be done to correct them?
5 Which manufacturers and types of equipment prove reliable for our work and which unreliable? This will help us to buy efficiently in the future.
6 Which maintenance men are good at their jobs and which need further help and training?
7 How large should the maintenance staff be to deal efficiently with the work at minimum cost?
8 What maintenance equipment should be purchased?

We can reduce the foregoing to the information we must collect. Thus:

1 When an item of equipment fails, we can ask:

a How did the operator know it had failed?

b Was it a complete failure or did it work to some extent, and if so what did it fail to do? Did it cause a safety hazard?

c Was the failure caused by:

(i) Overloading.

(ii) Misuse, neglect, damage, accident.

(iii) Arduous working conditions.

d What repair work was done, how long did it take and how many maintenance men were involved?

e What maintenance equipment was required and for how long?

f How long was the equipment out of service and what did it cost in:

(i) Operators' time.

(ii) Lost production if in industry or lost availability.

(iii) Spoilt work.

g Were there any symptoms by which the breakdown could have been predicted and avoided by scheduled maintenance?

2 A similar return when scheduled maintenance is done can be useful. Thus:

a What adjustments or replacements were made?

b What symptoms indicated that they were desirable?

c Were there any signs of overloading or misuse?

d How long did the scheduled maintenance take?

e Was it possible to do it at a time when the equipment was not required for use?

f What maintenance equipment was used and for how long?

The above looks a pretty formidable list but it attempts to foresee most possibilities. In any actual case, we only ask for information we require and which can be given with sufficient accuracy to be useful rather than misleading.

16.10. availability and costs

In section 4.3 we noted that availability was often more important to the user than reliability, and that it could be expressed by equation (4.3a) thus:

$$\text{Average availability} = \frac{\text{MTBF}}{\text{MTBF}+\text{MTTR}} \tag{4.3a}$$

At first sight we might think that a major objective of the maintenance engineer must be to ensure that every item of equipment is available for use as continuously as possible. Fortunately this is not necessarily so, for if it was it would largely make scheduled maintenance useless. After all, the basis of scheduled maintenance is that an item of equipment, which is operating acceptably, is deliberately made non-available while it is checked and any necessary corrections made. Thus scheduled maintenance nearly always *reduces* the average availability. Its attraction, however, is that we can choose a time when the equipment will not be required for use, whereas breakdowns often occur at most inconvenient times.

There are predictable times when many items of equipment are not required. Thus :

1 Television and radio equipment may only be required during broadcasting hours.
2 Space heating in buildings is usually only required during the winter.
3 Industrial equipment can often be serviced outside working hours.
4 Even in the armed forces, it will probably be preferable to have equipments out of use, say one at a time, rather than risk two or three breaking down together.

Whereas in industry a cost can be put on a breakdown, then it can be argued that to be effective scheduled maintenance must work out cheaper than risking breakdowns with their attendant loss of production. Overall then, we aim to have our plant and equipment available for use whenever required, but this is not necessarily all the time.

With complicated equipment of a new design it is not unusual to find that the time it spends out of service is quite considerable. Thus suppose that on average 30 per cent of our equipments of a particular type are not available for use due to some repair or adjustment. This means that only 70 per cent are available, and in order to have any given number actually in use we must purchase $100/70 = 1{\cdot}43$ times as many as we need. The extra 43 per cent of probably expensive equipment will in effect spend its time littering our maintenance workshops.

Where scheduled maintenance is used, it is often convenient to express the time between overhauls (TBO) rather than the MTBF. Thus, on aircraft, the old reciprocating engines driving propellers had a TBO of 1500 to 1800 hours. Modern jet engines are capable of achieving TBO's approaching 10,000 hours.

Finally, we must not forget to maintain the maintenance equipment. It is often complicated enough to need consideration in its own right.

PART FOUR **TESTING AND OTHER SOURCES OF RELIABILITY INFORMATION**

CHAPTER 17 SOURCES OF RELIABILITY INFORMATION

17.1 general sources of information

Throughout this book we have assumed that any information we need will be immediately available, but we have not fully considered its source. The following therefore is a list of some of the chief sources of such information, and we must then briefly consider the merits of each.

SOURCES OF RELIABILITY INFORMATION

1 Tests

a Demonstration and acceptance tests.

b Measurements to estimate θ, M σ and the constants of the Weibull distribution.

c Data from suppliers' tests.

2 Customer experience

a Market research.

b Customers' maintenance records.

c Returns for repair or replacement.

d Customer complaints.

3 Other sources of information

a Quality control department.

b Published data.

17.2. the advantages and disadvantages of reliability tests

Direct tests on parts, assemblies of parts or even occasionally on complete systems have the great advantage that we can control the conditions of test and record exactly what they are. In this sense they yield reliable data. Unfortunately tests tend to be both expensive and time consuming. Further, although we may know the precise conditions under which a test was done, we can never be sure that they exactly reproduce the treatment our products will receive in service.

Tests divide roughly into:

1 DEMONSTRATION AND ACCEPTANCE TESTS These are tests to decide whether a particular item or items can be accepted, and they are discussed in Chapter 19.

2 RELIABILITY MEASUREMENTS The designer who is considering specifying a particular item from a supplier will want actual figures on the parameters in which he is interested. Sometimes this information is available from that supplier, based on his own tests, and if we decide to use that product we can take advantage of his results. Care is necessary, however, because our supplier's test conditions may not really be comparable with the use conditions which we have in mind. Methods of making reliability measurements are discussed in Chapter 20.

17.3. customer experience

Information based on customers' experience has the advantage that it relates to actual use conditions, but unfortunately we seldom know exactly what these are. A typical situation arises when an item of equipment fails and is returned to the manufacturer who examines it, anxious to discover what went wrong. He then finds that, while there is no doubt that the item concerned has failed, there is no evidence whatever to establish how or why it happened. He is unable to reconstruct the circumstances just before failure. Probably the customer will not be able to help either, because he was not paying particular attention to that item of equipment at that moment. As far as he knows it was being operated normally.

It is sometimes assumed that information from customers is much cheaper than doing one's own tests, but this is not necessarily true. Such information usually concerns some sort of failure or malfunction, and so implies a measure of customer dissatisfaction. The cost of irritated customers may be more than we think, and also more than the cost of doing the equivalent tests.

Thus before we put a product on the market, we must do sufficient test and development work to ensure that its performance and reliability are worthy of the price we charge and that it will enhance our company's reputation. We shall not do this by rushing an immature product on to the market, and then in effect asking our customers to do our development tests as part of their everyday use of it.

Of course even the most conscientious of manufacturers will experience some failures in customers' hands. In addition, products which give satisfactory service will from time to time be returned for repair or adjustment and each occasion provides an opportunity to add to our knowledge of its reliability. Essential information should be recorded. Thus :

1 Serial number and type of equipment.
2 Why it failed.
3 What was done to correct it.

Unsatisfactory parts removed from an equipment can be subjected to a searching examination to determine the cause of failure.

Where customers keep maintenance records along the lines discussed in section 16.9, they may be prepared to make them available to the manufacturer and such an opportunity should not be missed.

17.4. analysis of failure data

As information collects, we shall want to analyse it for future use. Much of what we said in section 16.9 about the collection of maintenance information will also apply in principle to the collection of failure data. We shall only collect what could be useful to us, and we shall make sure that what is collected is fully analysed and used.

Both collection and analysis should be designed to suit the particular products and markets concerned, but the following gives a general idea how this can be done.

1 PRODUCT GROUPS

We shall want to know which particular products, or maybe groups of products, give our customers most trouble. Thus we might find that while our amplifiers give a lot of trouble our signal generators give very little. In concluding this, however, we must take account of the numbers of each sold. If so far we have sold 100,000 amplifiers and only 1000 signal generators, then we should expect far more failures on the amplifiers for the same failure rate. We must also remember, however, that any improvement on amplifiers will please a hundred times as many customers as a corresponding improvement on signal generators. This is not the only consideration. The amplifiers may be sold in the domestic market at a very competitive price, which implies that the reliability cannot possibly be outstanding, while the signal generator may be sold to the armed forces at a price which implies the highest possible reliability.

The use to which a product is put may also affect its reliability. Thus suppose we are studying failures which have occurred in delivery vans. We might find that vans used in large towns such as London have significantly more clutch and gear box failures than those used on main cross country roads.

2 PARTS

Analysis may show that particular parts or groups of parts cause a significant proportion of the failures. Thus one part may always give trouble regardless of the type of equipment in which it is used. One particular make may give trouble, while a similar part from another manufacturer is almost trouble free.

If we combine our analysis of failures by product and by individual parts, we may find that some parts give trouble when incorporated into certain products, but are almost trouble free when put into others. Presumably the part is treated differently in the two applications. The difference may be one of load stress, but it may be in things like ambient temperature, voltage surges, the use cycle, shock, vibration, etc. as discussed in Chapter 9.

3 MODE AND MECHANISM OF FAILURE

Having ascertained that a particular part causes a significant proportion of failures, we then have to consider what it is that it fails to do. This is called its *mode of failure*. Thus :

a An electrical relay fails to close a circuit when required.
b A car ignition key will not turn when inserted in the dashboard.

Next we shall need to analyse each mode of failure for the *mechanism* by which it failed. Thus the reason the relay does not close the circuit might be :

a The contacts are incorrectly positioned.
b The moving parts are jammed due to :
  (i) Corrosion.
  (ii) Damage.
c The magnetic field is too weak, due to :
  (i) Wide air gap.
  (ii) Low current due to terminal corrosion.

Notice that low current due to low supply voltage would lead us to look at the supply for the cause of failure.

When we have identified the mechanism of failure, we are probably only a short step from curing the trouble. Thus if corrosion of the relay is a significant cause of failure, we must either protect it or make it of corrosion resisting materials. Sometimes requirements are interrelated, so that we might find that corrosion resisting materials do not have the right magnetic properties or that a protective cover increases the size or weight unacceptably. This is a technical problem, but at least we know exactly where our difficulty lies.

17.5. the severity of defects and failures

In the foregoing we have spoken as though any defect resulting in a failure represents 'one standard unit of trouble', but clearly this is not necessarily the case. In section 1.4, we noted various ways in which failures could be subdivided. In addition to these, it is often useful to classify defects by their severity as follows.

A critical defect is one which is likely to result in hazardous or unsafe conditions for the people using, maintaining or depending on the product, or one which will cause the failure of a major item, such as a large radar system. Thus a defect which might result in the failure of the brakes of an omnibus would be classed as critical.

A major defect is one which although not critical will result in failure or will seriously reduce the utility of the product in which it is incorporated. Thus a defect which might cause an omnibus engine to either fail or perform very badly, would be classed as major.

A minor defect is one which although undesirable will not appreciably affect the use of the product concerned. Thus paint blemishes on an omnibus, if not too serious, might be classed as minor.

17.6. the annoyance caused by defects and failures

The annoyance caused to customers by defects and failures is not necessarily related to their severity. Clearly no one wants a car whose brakes and steering are unreliable, but we may find that a relatively minor defect, like the failure of the engine to start instantly or excessive engine noise, causes a quite disproportionate amount of annoyance to customers. In such cases the defect concerned must receive a correspondingly large amount of our attention.

17.7. the age of the system

The pattern of failures from a particular system may well relate to its age, probably expressed as the number of hours for which it has operated since it was new. This in turn may relate to the bath tub curve as shown in Fig. 17.7(a). Thus suppose we are concerned with the failure of two particular parts X and Y. X

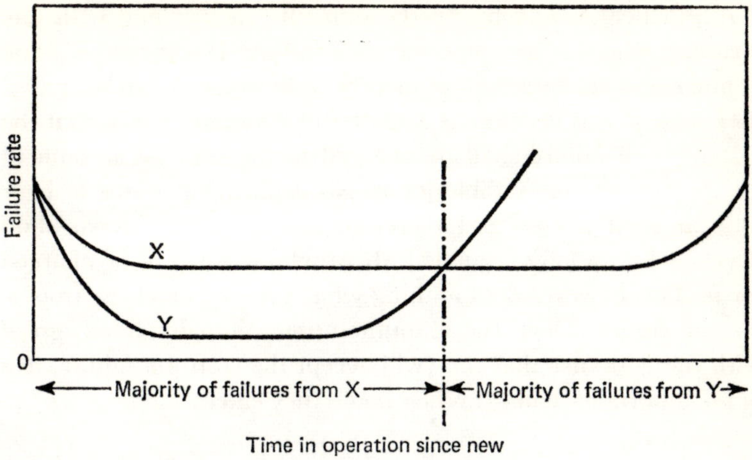

FIG 17.7(a). Failure patterns of two parts X and Y

has a long constant failure rate period, but a relatively high failure rate. Y, on the other hand, has a low failure rate during the constant period but wear out starts relatively early.

If these two parts are both included in one system, the pattern of failure will tend to be that early in its life failures will be mostly due to the high failure rate of X, but later on wear out failures from Y will dominate.

We shall be lucky if we can establish a relationship of this sort solely by analysing customer returns. Once we have a clue, however, we can do laboratory tests to establish the facts.

17.8. the area of the company responsible

Sometimes a very telling analysis can be deduced by sorting customer failures into the area of the manufacturing company responsible. Thus failures might be divided into those due to :

1 Design.
2 Manufacture.
3 Raw material or purchased items.
4 Damage in store or in transit.
5 Faulty installation.
6 Faulty customer operation.

A practical snag may be the difficulty in deciding with any certainty who is responsible for each failure. If someone is going to guess, the analysis may be of very little value. A further pitfall may arise if it is decided as a matter of company policy that the cost of each failure shall be charged to the expense account of the department responsible for it. No department wants to have to bear such a cost and therefore too often the investigation develops into a long argument about who is going to be charged for it. This however does nothing whatever to correct the trouble for the future. Once the manufacturing company have agreed with the customer that they will accept the cost of a failure, it is of no importance where in their books they enter it.

17.9. the Pareto distribution

In the foregoing sections, we have collected various groups of failures together. Often the analysis is clarified by plotting a Pareto distribution, as illustrated in Fig 17.9(a).

Vertically we plot the cumulative percentage of failures. Thus

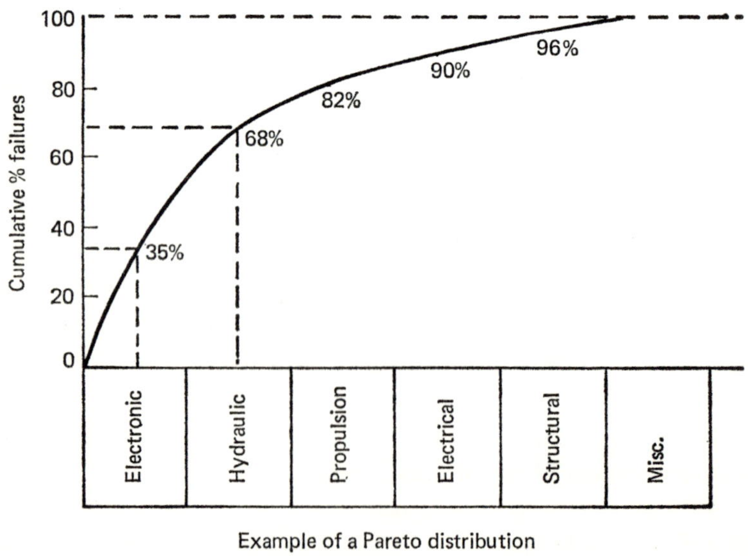

FIG 17.9(a). Example of a Pareto distribution

100 per cent represents all the failures observed over a convenient period of time. These are then divided into groups which are plotted horizontally starting with the largest group, and then going down in strict descending order. The power of the Pareto distribution lies in the fact that failures are almost never equally divided between the various possibilities. Thus, in the above example, 35 per cent of the failures were electronic and a further 33 per cent making 68 per cent in all, were hydraulic. In order therefore to reduce overall failures, we must concentrate on electronics followed by hydraulics. We should not for the moment worry about structural failures, unless of course they are critical defects, because they represent only 5 per cent of the total and, if we eliminated them completely, 95 per cent of our failures would still remain.

The next step in the analysis might be to draw another Pareto distribution, analysing the electronic failures by themselves, by parts, modes and mechanisms of failure. Similarly the vertical scale can be varied to suit the problem. Sometimes a money scale is more telling, because individual failures may vary considerably in what they cost.

17.10. setting up a reliability information system

As our collection and analysis of reliability information grows, we may find that there is more than can conveniently be handled by a few clerks, and we shall look for a more formal and efficient way of doing it. Our precise needs, however, will depend not only on the magnitude of the task but also upon whether we are in the position of manufacturer or user. Thus, within the armed forces, they naturally have a big interest in improving utilization and maintenance of equipment. A manufacturer is unlikely to have access to such detailed information about use and maintenance, but will probably add data from specific tests to that which he gets from customer returns.

We must decide precisely what data we intend to collect and how it is to be processed. It will need to be in a form which can be easily codified, since long explanations are difficult to tabulate.

Our discussions in section 16.9 are relevant here. If the information is to be computer analysed, then the whole system must be designed to collect data in the form required.

17.11. post mortem analyses

Sometimes after a failure has occurred it is necessary to carry out a 'post mortem' to try to find out which unit or block caused the whole system to fail.

Thus suppose we have a system which consists only of two units, A and B, connected in series and that:

$$\text{Reliability of unit A} = 0{\cdot}9$$
$$\text{Reliability of unit B} = 0{\cdot}8$$

Now both unit A and unit B must work for the system to be satisfactory. Therefore from equation (7.2b) we have:

$$\text{Reliability of system} = 0{\cdot}9 \times 0{\cdot}8 = 0{\cdot}72$$

Therefore:

$$\text{Probability system fails} = 1 - 0{\cdot}72 = 0{\cdot}28$$

If, after the system has failed, we have to investigate we know that B is twice as likely to be the cause as A.

$$\text{Thus probability A fails} = 1 - 0{\cdot}9 = 0{\cdot}1$$
$$\text{Thus probability B fails} = 1 - 0{\cdot}8 = 0{\cdot}2$$

Now A and B are quite independent of each other. The fact that A fails does not make B either more or less likely to fail, and so the statistician calls these *unconditional probabilities*.

Sometimes, however, probabilities are not independent of each other. Suppose we are concerned with both the reliability of a new machine and the safety of the operator who works it. We guard it well, set the operating routine and confidently predict that it is safe. An operator is set to work and within a few days has a serious accident when her long hair gets caught in the machine. We have fallen down, because we did not allow for the girl's long hair, and we conclude:

With short hair girl—machine safe.
With long hair girl—machine unsafe.

This is a conditional type of problem. The safety of the machine is not independent; it depends upon whether it is operated by a girl with long or short hair.

A similar situation arises if the reliability of a machine depends upon the skill of an operator. Suppose we have a machine which is 0·99 reliable when used by operator A, but only 0·90 reliable when used by a less skilled operator B. Further suppose that operator A works the machine for four times as many hours as operator B. Thus we have:

Probability machine fails, given that operator A works it
$= (1{\cdot}00-0{\cdot}99) = 0{\cdot}01$
Probability machine fails, given that operator B works it
$= (1{\cdot}00-0{\cdot}90) = 0{\cdot}10$
Probability operator A does in fact work it = 4 in 5 = 0·80
Probability operator B does in fact work it = 1 in 5 = 0·20
Probability machine fails and A is working it
$= 0{\cdot}01 \times 0{\cdot}80 = 0{\cdot}008$
Probability machine fails and B is working it
$= 0{\cdot}10 \times 0{\cdot}20 = 0{\cdot}020$
Probability machine fails (regardless of operator)
$= 0{\cdot}008+0{\cdot}020$
$= 0{\cdot}028$
Reliability of machine when operated as above
$= 1{\cdot}00-0{\cdot}028$
$= 0{\cdot}972$

Again we have a conditional type problem. If we let F be the probability the machine fails, then we can write:

Probability machine fails, given that A is operator = $P(F|A)$
Probability machine fails, given that B is operator = $P(F|B)$

$P(F|A)$ is mathematical notation and means the probability that F will occur, given that A has already occurred. This is called *conditional probability* because the probability of F depends upon whether A or B have previously occurred.

Consider now the reliability of a burglar alarm. We not only want it to sound when a burglar is present; we also want to be reasonably sure that it will not give a false alarm when there is no one there. Indeed we can set out four possible situations:

	Burglar	*Alarm*
1	present $= B_1$	sounds $= A_1$
2	absent $= B_0$	does not sound $= A_0$
3	present $= B_1$	does not sound $= A_0$
4	absent $= B_0$	sounds $= A_1$

Since (3) and (4) represent incorrect functioning of the alarm we can express its reliability as the probability of events (1) or (2). Again we are dealing with conditional probabilities, since the probability the alarm will sound depends we hope on whether or not a burglar is present.

Using the notation we have just discussed, we can write:

$$\begin{pmatrix}\text{Probability the}\\ \text{alarm sounds and}\\ \text{burglar is present}\end{pmatrix} = \begin{pmatrix}\text{Probability}\\ \text{burglar is}\\ \text{present}\end{pmatrix} \times \begin{pmatrix}\text{Probability that,}\\ \text{given burglar is pre-}\\ \text{sent, alarm will sound}\end{pmatrix}$$

writing this in symbols we get:

$$P(A_1B_1) = P(B_1) \times P(A_1 | B_1) \qquad (17.11a)$$

17.12. Bayes' theorem

Equation (17.11a) is fundamentally quite general, and expresses the probability that two events A and B will occur together when the probability that one will occur depends upon whether or not the other has occurred. Thus dropping the suffices we get:

$$P(AB) = P(B) \times P(A | B)$$

Therefore:

$$P(B) = \frac{P(AB)}{P(A | B)} \qquad (17.12a)$$

This is a simplified form of a theorem developed by Thomas Bayes in the eighteenth century. Bayes himself had some doubts about his reasoning, and what is now called Bayes' theorem was not published until after his death. Since then mathematicians have argued about how it should be applied.

The reliability engineer need not worry about the validity of equation (17.12a), but it is often difficult to put meaningful values to the symbols. Our discussion on the burglar alarm is sound in general terms, but as soon as we try to give values to such quantities as the 'probability a burglar will attempt to break in' we are in difficulties. Furthermore, the owner of the premises to be protected will probably attach much more importance to whether the alarm will sound if a burglar is present than to whether it may occasionally give a false alarm.

The following example shows just how careful we have to be with this type of problem. Suppose we develop a system which has two units in series, both of which are required to give the same performance. We have developed two alternative designs for these units and have designated them A and B. On test, A and B perform equally well and there appears to be nothing to suggest that one is in any way better than the other. We therefore construct three complete systems containing the following units:

System 1 has two type A units = (A+A)
System 2 has one type A unit plus one type B unit = (A+B)
System 3 has two type B units = (B+B)

We send these three units to an interested customer with a request for information about their performance and reliability in service. However, as so often seems to happen, our customer puts them to use without noting where each is sent.

Some time later, one of the systems comes to grief and is largely destroyed. Fortunately there is enough left to tell that one of the units in this system is type B. There is nothing left to tell us the identity of the unit twined with it, nor do we know where the other two systems are. The question we naturally ask ourselves is, 'What are the chances that the other unit in the failed system was also a type B, possibly indicating some inherent weakness in that design?' This problem is more tricky than it seems, because it has two obvious answers, both of which are incorrect. Thus:

1 We might argue that since only one of the three systems has a (B+B) arrangement, therefore the chances that it is the one

which has failed must be 1 in 3 or 0·333. This would be true if one of the three systems was selected completely at random, without our having any information about which it was. In our case, however, we know that one of the units in the failed system is a B, and that therefore it cannot be the one arranged (A+A).

2 Alternatively we might argue that since of the six units made three are type A and three type B, therefore the chances that any unit at random, such as the destroyed unit, is a B must be 0·50. This would be true if we did not know where any of the units were, but in fact we know that the failed system is not (A+A), and we also know where one B unit is. Hence two A units and one B unit are out of the reckoning.

By pursuing this line of reasoning, we can deduce the correct probability. Since we can account for two A units and one B unit, the destroyed unit must be one of the other three which are still unaccounted for, and these consist of two B units and one A. Hence the probability that the destroyed unit is type B is 2 in 3 or 0·666.

We could have deduced this result directly from Bayes' theorem as follows:

$$P(A\,|B)=\frac{P(AB)}{P(B)}$$

which becomes for this case:

$$P(B\,|B)=\frac{P(BB)}{P(B)}=\frac{\frac{1}{3}}{\frac{1}{2}}=\frac{2}{3}=0{\cdot}667$$

CHAPTER 18 RELIABILITY TESTING: GENERAL CONSIDERATIONS

18.1. practical difficulties

Since tests tend to be both expensive and difficult, we should not start one without full consideration of its probable cost, balanced against the value of the information we hope to obtain from it. We can judge the magnitude of our problem by considering the table of typical failure rates for parts given in section 6.7. This shows that a failure rate might be of the order of 0·01 per cent per 1000 hours, which means that on average the part will operate for 10,000,000 hours before a failure occurs. Now 10,000,000 hours is approximately 59,524 weeks or 1145 years, always assuming this we test for 24 hours a day and 7 days a week! Clearly this is impossible, but we can of course reduce the time by testing a group of parts together. Thus if we test 1000, we can accumulate 10,000,000 hours in a mere 1·145 years! But even this is only the time during which we would expect one failure. We should have to test for considerably longer to get a significant number of failures.

However, in many practical cases, the time available for testing is limited by the terms of the contract. If a complete system has to be designed, made and installed in, say, 3 years, the fraction of this available for development testing and generally proving the design will be measured in months at the most, and probably only in weeks.

A further limitation will be the availability of test rigs. If 1000 parts are to be put on test, in effect the test rig must provide 1000 independent tests. Allowing for the fact that a number of such tests are likely to be running at the same time, we can see that the cost of all the test rigs may be very high, and that even then there will only be a limited amount of room available on them.

Too often we find in practice that the design of a given test is decided more by what is possible, than by what should be done. Thus the sequence might be :

1 The number to be tested is set by the number of places available on the test rig.
2 The duration of the test is decided by the time available before the results must be ready.
3 In such circumstances the reliability engineer may be able to do no more than calculate the confidence, or maybe lack of confidence, which can be placed in the results.

18.2. types of test

Most reliability tests can be divided into one of two groups.

1 RELIABILITY DEMONSTRATION AND ACCEPTANCE TESTS In these tests our object is usually to show that the reliability of certain items is at least equal to, and therefore probably better than, some preset value. We do not however need to determine the precise value of their reliability. Thus a manufacturer, having produced certain equipment, may wish to demonstrate to a customer that its reliability is at least as good as he claims it to be. He therefore performs what he will call a *reliability demonstration test*. On the other hand, if the customer receives the equipment untested and then decides to do the tests himself, he will consider that he is doing a *reliability acceptance test*, to verify that the equipment satisfies the values for reliability given in the specification. Hence reliability acceptance tests and demonstration tests are fundamentally the same, but done from a different viewpoint. We shall discuss these tests in detail in Chapter 19.

2 RELIABILITY MEASUREMENTS Where mere demonstration is insufficient, then tests can be devised to estimate some parameter in figures. This might be the reliability itself, but in many cases it is the failure rate, MTBF or the constants of a Weibull distribution which are required. These tests are fully discussed in Chapter 20.

In addition to reliability proper, tests can be devised to check almost any aspect, for example:

1 MAINTAINABILITY For example the relative speeds with which the same repair can be carried out on two alternative designs. If the precise method for doing a repair is known, the time it will take can be worked out using one of the established predetermined motion time systems such as MTM or SPMTS. With these, the time can be found before anyone has actually done the job.

2 OPERABILITY For example the frequency of operator mistakes or the time required to do a job.

3 PRODUCT QUALITY This is the field of quality control, but there is no sharp dividing line. Often the better the quality of an article as made, the better the subsequent reliability.

The type of test listed immediately above can often be carried out in a relatively short time, and therefore might be preferred to a reliability type test if the required information could be obtained from it.

18.3. testing elements, units, blocks or complete systems

So far we have implied that tests will be made on individual parts, such as resistors, capacitors or maybe mechanical linkages. Clearly this need not be so, and we can test anything we like. However, difficulties and costs multiply as we scale up. If a 1000 parts are expensive to test, 1000 elements or units will be even more so and in general a test on 1000 complete systems will not be practicable unless they are very small.

Often the only real test on a complete system is that carried out by the customer, in the course of normal operation. Even the manufacturer may only do it by parts as explained in section 19.6.

18.4. accelerated tests

One way to reduce the very long test times which are often

necessary is to use an accelerated test. The idea is that the normal life cycle of the item concerned should be compressed into a very much shorter time. It takes two main forms.

1 *Tests done under normal use conditions, except that what would in practice be intermittent use becomes either continuous use or something approaching it.* Consider, for example, the ordinary cheap camera which people buy for their holiday photographs. The manufacturer wishing to life test these might argue that the average number of photographs taken per year seldom exceeds, say, 100 and most people expect say 10 years' use. Hence the acceptable life of such a camera is about 1000 photographs. Now the shutter mechanism of a camera can easily be given 1000 operations in the course of an hour or so, and after that it can be run on until it fails just to find out how long it will last. Hence a life test to destruction, representing more than 10 years of life can be completed in a few hours.

A somewhat similar test can be devised to test the wearing properties of a carpet. Thus a sample of the carpet might be installed in the foyer of a theatre, or even in a railway station. The number of people walking on it would vastly exceed what it would get in a home, and so its wearing qualities would be quickly determined. Furthermore, the number of tickets issued by the theatre or railway station would give an estimate of the number of people who had walked on it.

In both of the above examples, although the life of the article was compressed into a short time, the conditions of use were approximately normal and the results will probably be reasonably reliable.

2 *Tests where the normal conditions of use are intensified.* Suppose we want to test a batch of resistors or capacitors. Calculation along the lines discussed in section 18.1 shows us that at normal conditions our test will take years to complete. We therefore decide to increase the failure rate artificially, by increasing both the test voltage and the ambient temperature. In this way we argue, we can detect potential failures in a few weeks instead of years. Unfortunately, however, this type of test must be approached with considerable caution. Suppose we submit three

batches of parts from different manufacturers to our accelerated test and find that batch C fails long before batches A and B. Can we safely conclude that batch C will be worse in customer use than A or B? Until we have carried out tests to correlate our test conditions with practice we can make no such prediction. It may well be that C will prove equally good in practice.

18.5. catastrophic and degradation failures

It is useful to distinguish between catastrophic and degradation failures, as defined in section 1.4. The catastrophic failure is both sudden and complete, and in general there is no way to predict it. In tests, therefore, we just have to wait until one occurs, and this may take a very long time. The degradation failure however takes place gradually, and we may be able to observe some deterioration long before actual failure takes place.

Thus the forward resistance of some selenium rectifiers tends to increase roughly as indicated in Fig. 18.5(a). If a life test is carried out, it is possible to distinguish very early on between rectifier A which is unacceptable and rectifier B which is satisfactory. There are dangers, though, because rectifier C also appears to be satisfactory at the start but eventually rises past the failure limit. This test is further complicated because in order to

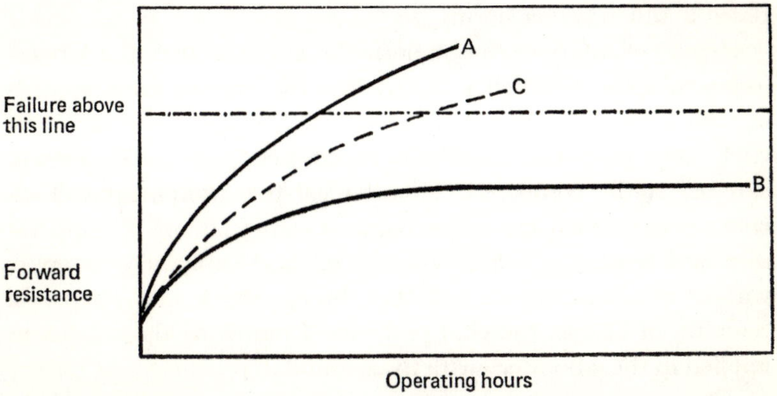

FIG 18.5(a). Increase in the forward resistance of a selenium rectifier with time in service

get results in any reasonable sort of time it must be done at accelerated conditions. Nevertheless degradation failures often offer easier test possibilities than do catastrophic failures.

Sometimes the potential life of an item can be estimated by a near instantaneous test. Thus the strength of a weld or soldered joint would probably be measured by the force required to pull it apart, rather than by a protracted life test to find out how long before it failed.

18.6. environmental tests

In general we do not want to know whether an item will work satisfactorily in a clean, vibration and shock free, temperature controlled laboratory. We want to satisfy ourselves that it will be all right on the job, when subjected to the hazards discussed in Chapter 9. Therefore many life tests will be done under simulated environmental conditions. They can of course be done in the actual working environment, and we may think that nothing could be better than the real thing. Thus painted samples are sometimes left out in the open, exposed to the weather for months, so that their deterioration can be observed. Control of such a test is difficult however. Thus if two equipments are each exposed to tropical storms, and A fails while B survives, we have to decide whether A failed because it was inferior, or because it was exposed to much fiercer storms.

Hence reliability engineers normally install in their test laboratories equipment which will simulate the various environments in which they are interested. Their test conditions are then precisely known, and the reliability of different types of equipment can be validly compared. Thus typical test equipment will include ovens which can apply various combinations of temperature and humidity, refrigeration plant and equipment to apply various combinations of vibration, bump, shock, etc. There still remains, of course, the vital problem of matching the conditions applied in the laboratory with those found in practice.

CHAPTER 19 RELIABILITY DEMONSTRATION AND ACCEPTANCE TESTS

19.1. basic considerations

Suppose we want to demonstrate that one of our products is as reliable as we say it is or that a source of supply is as reliable as the supplier claims. We must first decide what test is to be applied, taking account of anything specified in the contract, the use conditions and all that we said in Chapter 18. We must then set out the precise conditions of the test, including electrical and/or mechanical loadings, environmental conditions and duration.

Next we must decide how many items to test. If we test only one, it tells us very little. Should it fail, we do not know whether it is typical of its type or the odd bad one in perhaps thousands of good ones. Yet because tests are expensive we want to test as few as possible, so we must consider exactly what we learn from each test. Suppose we have to test a supply of parts whose actual reliability, unknown to us, is 0·90 or 90 per cent.

If we test 1 part only:

Probability it survives $= 0{\cdot}90 = r$

Probability it fails $= 0{\cdot}10 = f$

Probability it either survives or fails $= 1{\cdot}00 = (r+f)^1$

If we test 2 parts:

It is easiest to think of this as a test on one part, followed by a test on a second.

Probability both survive $= 0{\cdot}90 \times 0{\cdot}90 = 0{\cdot}81 = r^2$

Probability 1st survives and 2nd fails $= 0{\cdot}90 \times 0{\cdot}10 = 0{\cdot}09 = rf$

Probability 1st fails and 2nd survives $= 0{\cdot}10 \times 0{\cdot}90 = 0{\cdot}09 = fr$

Probability both fail $= 0{\cdot}10 \times 0{\cdot}10 = 0{\cdot}01 = f^2$

Total $= 1{\cdot}00 = (r^2 + 2rf + f^2)$

$= (r + f)^2$

In order to make our argument quite general, suppose that parts have a reliability or probability of survival = r, and an unreliability or probability of failure = f. Since each part must either survive or fail,

$$r + f = 1{\cdot}00 \text{ as equation (3.2d)}$$

If in the above calculation, we now add these general terms, as shown on the right hand side, then we get :

For a test on 1 part $(r+f)^1$

For a test on 2 parts $(r+f)^2$

and if the reader cares to work this out for any other number of parts, say n, he will always get :

$$(r+f)^n \tag{19.1a}$$

Mathematically minded readers will recognize equation (19.1a) as the binomial distribution. Thus suppose we have to test a sample of 10 parts. By expanding $(r+f)^{10}$, we can deduce the probability of getting any combination of survivors and failures as follows :

$$(r+f)^{10} = r^{10} + 10r^9f + 45r^8f^2 + 120r^7f^3 + \ldots + f^{10}$$

Therefore in the case where the true reliability of the parts is 0·9, we get :

Probability of 0 failures in 10 $= r^{10} = (0{\cdot}9)^{10} = 0{\cdot}35$ (19·1a)

Probability of 1 failure in 10 $= 10r^9f = 10 \times 0{\cdot}9^9 \times 0{\cdot}1 = 0{\cdot}39$

Probability of 2 failures in 10 $= 45r^8f^2 = 45 \times 0{\cdot}9^8 \times 0{\cdot}1^2 = 0{\cdot}19$

Probability of 3 failures in 10 $= 120r^7f^3 = 120 \times 0{\cdot}9^7 \times 0{\cdot}1^3 = 0{\cdot}06$

So far we have only considered the case where the actual reliability of the parts is 0·9. Since their reliability could be any value, we will also calculate the corresponding probabilities where $r = 0{\cdot}8$, 0·7 and 0·6 and this gives the following table.

Number of parts which fail, in a sample of 10	Probability that the number of failures in the left hand column will in fact occur, if the actual reliability of the parts is as follows			
	0·90	0·80	0·70	0·60
0	0·35	0·11	0·03	0·005
1	0·39	0·27	0·12	0·04
2	0·19	0·30	0·23	0·12
3	0·06	0·20	0·27	0·21
Total probabilities of 0, 1, 2 or 3 failures	0·99	0·88	0·65	0·375

Table 19.1a

19.2. the operating characteristic

We must now reduce the above figures to something convenient for practical use. Consider the case in which we test 10 parts, and they all survive, with 0 failures. We want to know what confidence we can place in the reliability of the supply of parts from which that sample was selected. To deduce this we plot what is called an *operating characteristic*, as shown in Fig. 19.2(a), using the row of figures above which is opposite to zero failures.

Horizontally we set out possible values of the actual reliability of the supply of parts from which our sample of 10 came, and vertically we put the probability that there would be 0 failures out of 10 for each value of actual reliability. Notice that for a reliability of 1·00, the probability of 0 failures must also be 1·00 and this is true for every operating characteristic. As the reliability of the parts reduces so the probability of getting zero failures in a sample of 10 reduces.

Suppose that the parts we are testing have an actual reliability of 0·80. Our operating characteristic then tells us that the probability of 0 failures in 10 is only 0·11 or 11 per cent, so therefore there must be a probability of (100–11) = 89 per cent that we

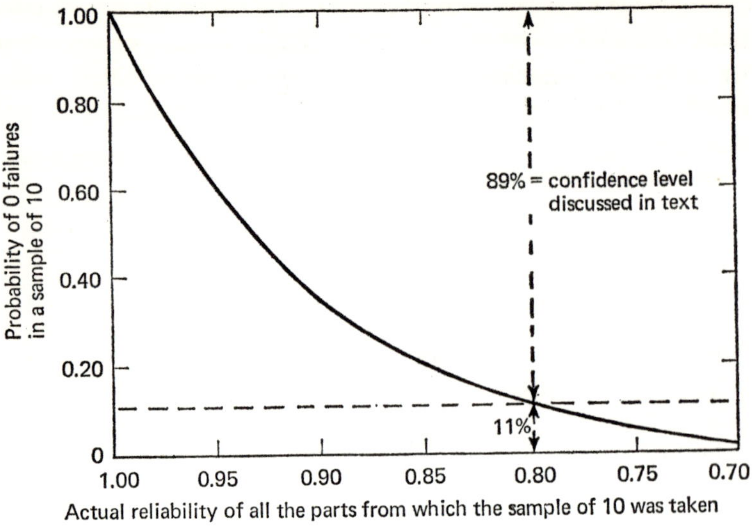

FIG 19.2(a). Example of an operating characteristic

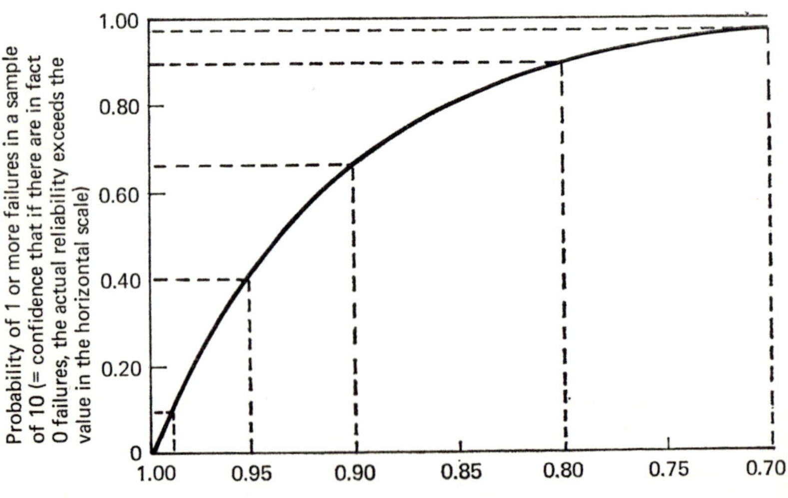

FIG 19.2(b). Example of an operating characteristic showing the probability of 1 or more failures

shall get 1 or more failures. This means that if we test 10 parts from a supply whose reliability is unknown, and get 0 failures, we know that there is only an 11 per cent chance that the reliability is as low as 0·80, and we are 89 per cent confident that it must be better than 0·80.

Following quality control practice, it is usual for operating characteristics to be plotted as in Fig. 19.2(a), but it is in fact easier to appreciate their reliability implications if they are plotted the other way up, as in Fig. 19.2(b). Vertically we now have the probability of 1 or more failures in our sample of 10. Thus when the actual reliability was 0·80, the probability of zero failures was 0·11 or 11 per cent and the probability of 1 or more failures was 89 per cent. Thus we are now plotting vertically our confidence that the actual reliability is better than the value given horizontally. For this particular example, where 10 parts are tested and 0 failures occur, we have a confidence of :

97% that the actual reliability is better than 0·70
89% that the actual reliability is better than 0·80
65% that the actual reliability is better than 0·90
40% that the actual reliability is better than 0·95
10% that the actual reliability is better than 0·99

and of course zero confidence that the actual reliability is better than 1·00.

19.3. interpretation where some failures occur

Suppose however that some failures do occur in our sample of 10. We can still use the table at the end of section 19.1. If as before the actual reliability is 0·80, then the table tells us that the probability of 0 failures is 0·11 and of 1 failure is 0·27. Since $0{\cdot}11+0{\cdot}27 = 0{\cdot}38$, the probability of 2 or more failures must be $1{\cdot}00-0{\cdot}38 = 0{\cdot}62$. Thus, if we get 1 failure in 10, we are 62 per cent confident that the actual reliability must exceed 0·80.

If we recalculate Table 19.1(a) on these lines, we get the following.

Number of parts which fail in a sample of 10	Probability that there will be *more* failures than is shown in the left hand column, if the actual reliability of the parts is as follows			
	0·90	0·80	0·70	0·60
0	0·65	0·89	0·97	0·995
1	0·26	0·62	0·85	0·95
2	0·07	0·32	0·62	0·83
3	0·01	0·12	0·35	0·62

Replotting our operating characteristics in the style of Fig. 19.2(b), we get Fig. 19.3(a).

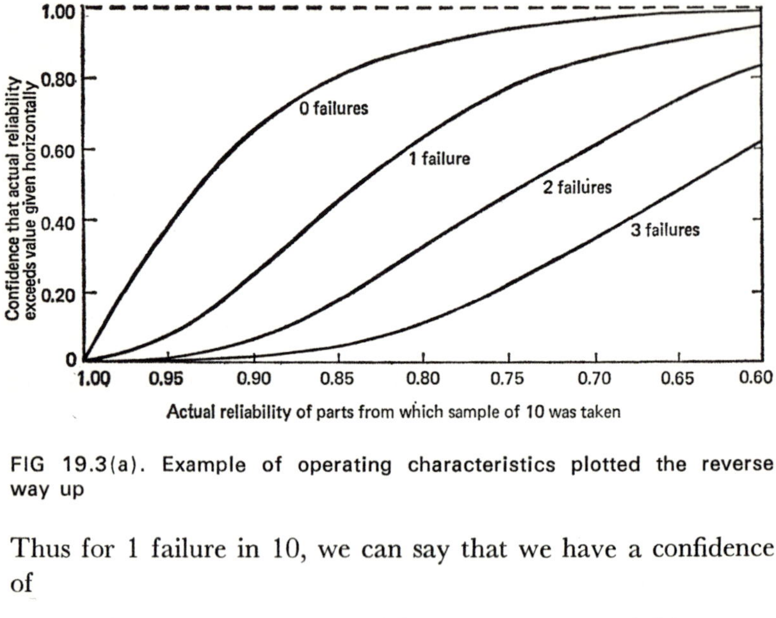

FIG 19.3(a). Example of operating characteristics plotted the reverse way up

Thus for 1 failure in 10, we can say that we have a confidence of

95% that the actual reliability is better than 0·60
85% that the actual reliability is better than 0·70
62% that the actual reliability is better than 0·80
26% that the actual reliability is better than 0·90

Since operating characteristics are usually drawn as shown in Fig. 19.2(a), the reader should familiarize himself with this form so that he can read confidence levels from them.

19.4. the choice of an operating characteristic and confidence levels

If we decide how many items to test and how many failures we will permit, then we can calculate the operating characteristic as we discussed in sections 19.1 and 19.2. Conversely, if we select a suitable operating characteristic, we can deduce the sample size and the allowed number of failures associated with it. Our object must obviously be to choose an operating characteristic

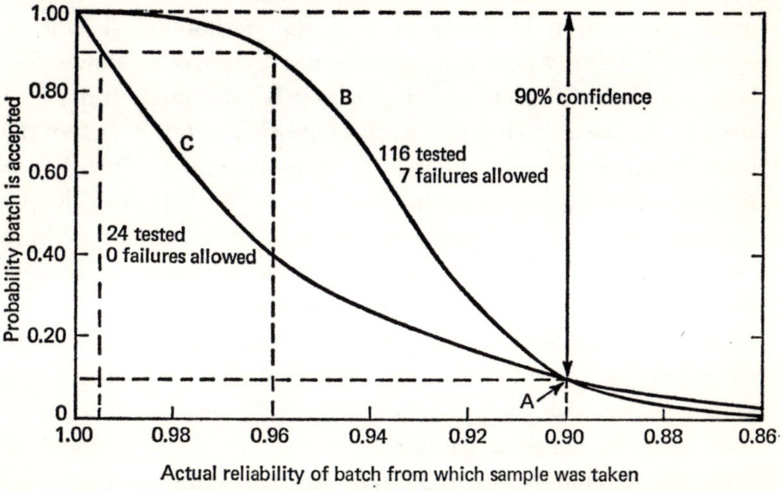

FIG 19.4(a). The effect on the shape of an operating characteristic when the sample size and allowed number of failures are increased

which will accept what we consider to be satisfactory reliability and reject what we decide is unsatisfactory.

Now if we increase the number of failures which are allowed, the operating characteristic moves to the right, flattening a bit as it goes. This is illustrated in Fig. 19.3(a). If we increase the number tested, it becomes steeper. The combined effect of increasing both the number tested and the allowed number of failures is shown in Fig. 19.4(a). Both the operating characteristics illustrated can be used to demonstrate a reliability of 0·90 with a confidence of 90 per cent (i.e. when the actual reliability is 0·90, the probability of acceptance is $(1-0{\cdot}90) = 0{\cdot}10$). Hence the curves meet at point A, and for reliabilities below 0·90 there is little to choose between them. However, above a reliability of 0·90 they are quite different, and since one requires a sample of only 24, while the other requires 116, we must consider what we get for testing so many more.

Now when we decide to demonstrate a reliability of 0·90 at a confidence of 90 per cent, we really mean that we intend to be 90 per cent certain that the actual reliability of the batch exceeds 0·90, but we have not said by how much. Suppose for example, that the actual reliability of the batch was 0·96. If we test 116 items, then curve B shows us that the probability it will be accepted is 0·90, whereas if we only test 24 curve C shows that there is only a 0·39 chance that it will be accepted. Indeed on curve C the actual reliability must be as high as 0·996 before the batch has a 90 per cent chance of being accepted. This then is the big difference between the two curves. Both have a 90 per cent probability of rejecting a reliability of 0·90, but curve C has a serious risk of rejecting appreciably higher reliabilities as well, whereas with curve B this risk is much smaller.

We must also decide the confidence level at which to carry out a test, and our first reaction will probably be to make this as high as possible. Suppose however that we want to demonstrate with 90 per cent confidence that a certain large batch of items has a reliability of 0·99. We can calculate the minimum number we must test from equation (19.1b) since:

$$r^n = (0{\cdot}99)^n = (1-0{\cdot}90) = 0{\cdot}10$$

Therefore:

$$n = 230$$

Now a reliability of 0·99 is not particularly high if we are dealing with individual parts, but it might apply to assemblies. However, to test 230 complete assemblies, possibly for many hours, might well be expensive. Further we are hoping for 0 failures, so if we get any at all we might well have to extend the test before we can be satisfied. We are then going to report that we are 90 per cent confident that the reliability of the items concerned is in fact better than 0·99.

Now if we wrongly inform a designer or supplier that he has

failed to achieve the required reliability, he may spend a lot of money, time and effort producing a much more expensive alternative. Consider therefore what will happen if we reduce our confidence level, and as an illustration we will recalculate the sample size for the same reliability of 0·99, but with a confidence level of only 60 per cent.

Therefore:

$$r^n = (0{\cdot}99)^n = (1-0{\cdot}60) = 0{\cdot}40$$

Therefore:

$$n = 91$$

If this is acceptable, we gain in at least two ways:

1 The number to be tested is only 91 instead of 230.
2 The more we test, the greater the risk we shall incur a failure. Hence if we only test 91 instead of 229, there is a much greater chance that they will pass the test.

Against this, we must consider carefully whether a confidence of 60 per cent instead of 90 per cent will be adequate. Now 60 per cent confidence means that there is a 60 per cent chance that the true reliability is better than we say, and a 40 per cent risk that it is worse. We still have a better than even chance of being on the 'safe' side, so while it is true that this implies some risk of being wrong, we may consider the risk to be justified because it reduces the number to be tested and therefore the cost of the test. This is especially so when the items under test come from a supplier who has proved his reliability in the past.

19.5. the effect of a constant failure rate

Where the failure rate is known to be constant, a further simplification is possible. Suppose we need to test 50 parts for 1000 hours each, but cannot afford to wait for 1000 hours, which is roughly 6 weeks of continuous testing. If the failure rate is constant, it will be exactly the same during the second 500 hours of the test as it was during the first 500 hours. Therefore if we test *two* parts each for 500 hours, we shall get exactly the same results as if we tested one part for 1000 hours. Thus it is only the total time that all the parts are collectively on test that is important, and within reason we can divide this among as many parts as we wish.

In our example we have to test 50 parts for 1000 hours each. Therefore:

$$\text{Total test time of all the parts} = 50 \times 1000$$
$$= 50{,}000 \text{ hours}$$

We could achieve this with:

5 parts for 10,000 hours each
50 parts for 1000 hours each
500 parts for 100 hours each or even
5000 parts for 10 hours each

Hence our test is really specified by the total number of hours the parts are collectively on test and the allowed number of failures during this time. Within reason we can arrange this total time to suit ourselves, although we shall be wise to avoid extremes. Thus if the number of parts on test is very small, there is a risk that they may not be a representative sample of the group whose reliability is to be demonstrated. If the number on test is unduly large, the time on test will be correspondingly short, and this again may lead to non-typical failure rates being observed.

Now if the MTBF was 5000 hours, we should expect on average 1 failure every 5000 hours for each part on test. Therefore since the test is to run for 50,000 hours, we should expect $50{,}000/5000 = 10$ failures, although the actual number observed might vary a bit from this. If we express the total test time in units of the MTBF, then the expected number of failures is the same as the number of MTBFs. Thus, as the above shows, since the test is to last for 10 MTBFs we shall expect 10 failures. In most practical cases, however, the number of test hours used for acceptance sampling will be insufficient to permit any accurate estimation of either the failure rate or the mean time between failures. The method for doing this is described in the next chapter.

It is possible to express the operating characteristic in MTBFs as shown in Fig. 19.5(a). Thus we can say that we wish to demonstrate with 90 per cent confidence that the MTBF of a given batch is better than some value θ_1, and this means that the operating characteristic is to have only a 10 per cent probability of accepting a batch whose MTBF is exactly θ_1. In order to denote the steepness of the operating characteristic, we must then set another and larger MTBF, say θ_0, at which we wish to be

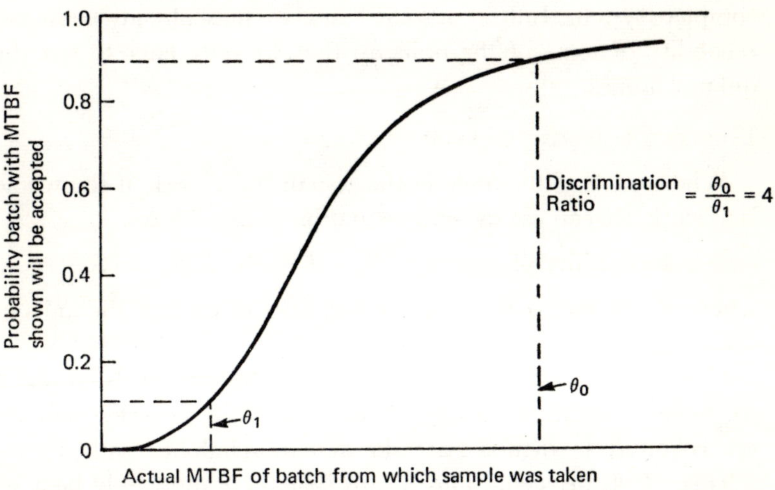

FIG 19.5(a). Example of an operating characteristic plotted in terms of the MTBF

fairly certain to accept the batch. This degree of certainty might by 90 per cent.

If θ_0 is only a little above θ_1, the operating characteristic will have to be very steep so that it can discriminate between the two. Conversely, if θ_0 is much greater than θ_1, only a flat operating characteristic is needed. Hence the ratio θ_0/θ_1 is called the *discrimination* ratio. The greater its value the flatter the operating characteristic, whereas when its value approaches 1·0 a very steep characteristic is necessary.

When MTBFs are used in this way, the sample size is expressed as the total test time required, regardless of the number of items between which it is shared. A preset number of failures is then permitted.

19.6. testing a system by parts

It is unlikely to be either economic or practicable to test a complete system of any size on the basis of putting any appreciable number on test. Hence systems are normally tested by parts. In principle we break the system down into convenient units or parts, test each separately and then calculate the reliability of the complete system. Indeed in some cases, we may already know the reliability of some of the units, so that we only have to test the unknown ones.

1 UNITS OR PARTS IN SERIES

Where each of the units in the system must work, if the whole is to work, we can use the series equation (7.2b). Thus:

$$\text{Reliability of system} = R_s = R_1 \times R_2 \times R_3 \times \ldots \times R_j$$

where R_1, R_2 up to R_j are the reliabilities of each of the units in turn.

Consider next what happens to our confidence levels. Suppose we have a system consisting of only two units in series, and we are required to demonstrate the system reliability with a confidence of 90 per cent. This means that there must only be a 10 per cent risk that the true reliability is in fact lower than the figure demonstrated. Now we shall only declare the whole system to be satisfactory if both units pass our tests. The situation is similar to that in equation (7.2b) above, and we can write:

$$\begin{pmatrix}\text{Probability system}\\ \text{accepted}\end{pmatrix} = \begin{pmatrix}\text{Probability 1st}\\ \text{unit is accepted}\end{pmatrix} \times \begin{pmatrix}\text{Probability 2nd}\\ \text{unit is accepted}\end{pmatrix}$$

$$P_s = P_1 \times P_2$$

But our confidence level is in fact the probability that a unit will be rejected. If therefore we denote the confidence level by C, we get:

$$P_s = (1 - C_s) = (1 - C_1) \times (1 - C_2)$$

In the simplest case, we shall allocate the same confidence level to each unit, denoted say by C_{12}.

$$P_s = (1 - C_s) \quad = (1 - C_{12})^2$$

$$P_s = (1-0{\cdot}90) = (1-C_{12})^2$$
$$\sqrt{0{\cdot}10} = (1-C_{12})$$
$$C_{12} = 0{\cdot}68$$

Hence the confidence level of 0·68, at which we can accept each unit, is actually *lower* than that required for the whole system of 0·90.

2 UNITS OR PARTS IN PARALLEL

Suppose now that we have to test a system of two or more units in parallel. This means, as we said in Chapter 8, that we shall not reject the whole system unless all the units in parallel fail. Thus for two units, we have from equation (8.2a);

$$\text{Probability both units fail} = F_s = F_1 \times F_2$$

But the probability of rejection is the same as the confidence level:

$$C_s = C_1 \times C_2$$

If as before $C_s = 0{\cdot}90$

and $C_1 = C_2 = C_{12}$, say

Therefore: $0{\cdot}90 = C_{12}$

$$C_{12} = 0{\cdot}95$$

Hence the confidence levels for units or parts in parallel must exceed that of the whole system.

19.7. practical considerations in reliability demonstration

Many practical considerations may influence the number of items we actually test, and the way we do or do not break a system into parts or units. In sections 19.1, 19.2 and 19.3, we showed that the theoretical sample size is determined in principle from the terms of the binomial distribution, i.e. from

$$(r+f)^n \tag{19.1a}$$

In practice, however, the binomial distribution is cumbersome unless the sample size is very small, and it is therefore usual to utilize the Poisson distribution instead. This introduces an approximation, but the error is usually negligible. Since tables giving values of the Poisson distribution are readily available, we do not have to do much working out.

Further economies in testing are frequently introduced, the most important of which arises when there is only a limited number of items from which the sample can be drawn. In section 19.1 we assumed that, since our items had an overall reliability of 0·90, every item we tested had a 0·90 chance of surviving and a 0·10 chance of failing. To take an extreme case, suppose that only 10 items are available for test and that, to match our assumption of 0·90 reliability, it contains 9 which will survive the test and 1 which will fail. If we could test them one at a time, we would get:

$$\text{Probability 1st fails} = \frac{1}{10} = 0{\cdot}100$$

If however it survives, then when we draw the second sample we get:

$$\text{Probability 2nd fails} = \frac{1}{9} = 0{\cdot}111$$

and continuing, always assuming that the defective item is not selected, we get:

$$\text{Probability 3rd fails} = \frac{1}{8} = 0{\cdot}125$$

$$\text{Probability 4th fails} = \frac{1}{7} = 0{\cdot}143$$

$$\text{Probability 5th fails} = \frac{1}{6} = 0{\cdot}167$$

$$\text{Probability 6th fails} = \frac{1}{5} = 0{\cdot}200$$

$$\text{Probability 7th fails} = \frac{1}{4} = 0{\cdot}250$$

$$\text{Probability 8th fails} = \frac{1}{3} = 0{\cdot}333$$

$$\text{Probability 9th fails} = \frac{1}{2} = 0{\cdot}500$$

$$\text{Probability 10th fails} = \frac{1}{1} = 1{\cdot}000.$$

Clearly, if the defective item is not one of the first nine, then it absolutely must be the tenth.

This means that if the number of items available for test is limited, the number tested can be reduced. Some care is necessary, however, and much depends on just what is available. Thus suppose that a small number of a new type of part is available from a supplier and we want to test them to decide whether the supply as a whole is likely to be satisfactory. Everything then depends upon the extent to which the parts available to us are really typical of the bulk supplies we shall receive later. If they are not typical, the test is probably fallacious from the start.

If the number of items available is very limited, this in itself may set the sample size since we cannot test more than we have. Other considerations include:

1 The availability of test rigs.
2 Whether every part tested is destroyed.
3 Cost of each test.
4 The time available for the tests.

It may be worthwhile to re-examine the choice of the reliability which has to be demonstrated. Thus suppose we have to demonstrate a reliability of 0·99 with a confidence of 90 per cent. It is worth considering how the figures 0·99 and 90 per cent were derived. In some cases there may be a very sound reason for them, but it is possible that they are little more than 'guesstimates,' maybe someone's idea of a target for which to aim. As we have seen, the confidence level in particular can be subject to review and occasionally very low values may be forced upon us.

We must of course take note of any contractual requirements. If we have a contract in which we have agreed to demonstrate a reliability of 0·99 at 90 per cent confidence, then we can only change this with the agreement of our customer.

19.8. selecting a suitable reliability demonstration test

In order to select a suitable reliability demonstration test, we have in principle to set four pieces of information, as illustrated in Fig. 19.8(a). These in turn specify the operating characteristic. The first two are:

1 The value of reliability to be demonstrated.
2 The confidence level at which this demonstration is to be made.

If we decide upon a confidence level of 90 per cent, this means that we wish to be 90 per cent certain of rejecting that value of reliability, with only a 10 per cent risk of accepting it. If nevertheless we do accept a batch with this reliability or worse, merely because the sample happened to be untypically good, it is to the consumer's disadvantage. Hence following quality control practice, this 10 per cent is called the *consumer's risk*.

The quality engineer uses the term *lot tolerance percentage defective* (LTPD) to denote the percentage of defectives in batches which he wishes to reject. Hence the reliability to be demonstrated and the LTPD are the same quantities expressed in different ways. For example, a batch with a reliability of 0·90 contains 0·10 or 10 per cent of items which will fail under test.

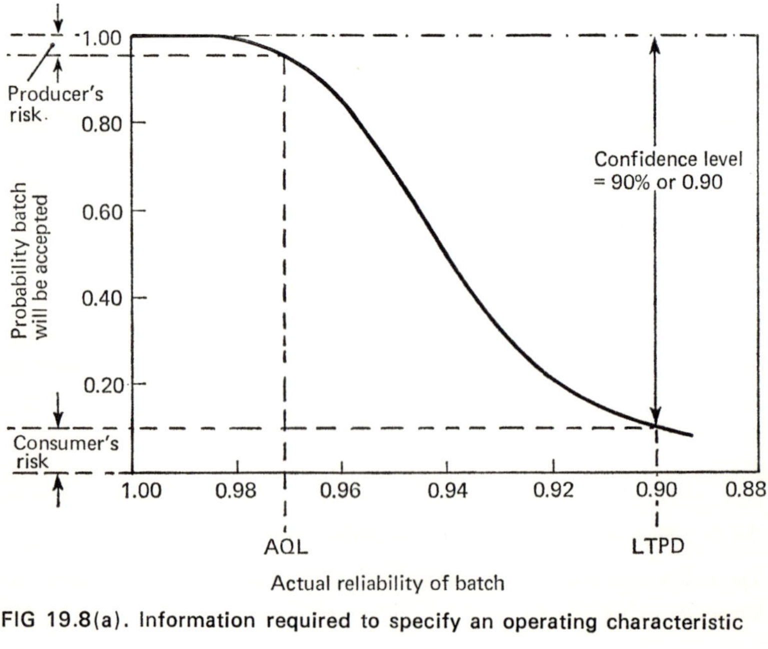

FIG 19.8(a). Information required to specify an operating characteristic

Thus the demonstration of a reliability of 0·90 at 90 per cent confidence corresponds exactly to inspecting the quality of a batch using a LTPD of 10 per cent and a consumer's risk of 10 per cent. Because of this exact parallel, sampling tables prepared for quality control can be used for reliability tests.

The remaining two pieces of information to be set for our reliability test are:

3 The value of reliability at which we wish to be fairly certain to accept the batch.
4 The risk we are prepared to take of rejecting a batch with that reliability.

If we decide to be 95 per cent certain of accepting the batch, the risk that we shall reject it and return it to the producer is 5 per cent or 0·05. Hence, as shown in Fig. 19.8(a), this is called the *producer's risk*. The value of reliability which we wish to accept might in quality control terms be called the process average and associated with a producer's risk of 0·10, but more likely it would be called the *acceptable quality level* (AQL) in which case it would be associated with a producer's risk of about 0·95.

With so many practical considerations there is a lot to be said for devising each sampling plan to suit an individual case. In order to give the reader some guidance, however, a table which applies to large batches has been inserted at the end of the book. Horizontally at the top, this sets out a range of reliabilities which we might want to demonstrate, and vertically on the left a range of allowed number of failures. Thus suppose we wish to demonstrate a reliability of 0·95 at 90 per cent confidence. The table gives us a range of sample sizes n, together with the corresponding allowed number of failures. We can choose any of these sampling plans. If we only test 45 with 0 failures allowed, we shall have a very flat operating characteristic, but as the sample gets larger so the characteristic gets steeper. The column headed $RPA = 0{\cdot}90$, therefore, gives the value of reliability which the sampling plan is 90 per cent certain to accept. The larger the sample, the lower this is, and the better the discrimination as explained in section 19.4.

Sampling plans at other confidence levels and reliabilities can be extracted from either of the following, even though both were primarily prepared for quality control use.

1 Defence specification DEF–131–A entitled, *Sampling Procedures and Tables for Inspection by Attributes*, which can be obtained from Her Majesty's Stationery Office.

2 British Standard 9001 : 1967 entitled, *Specification for Sampling Procedures and Tables for Inspection by Attributes for Electronic Parts of Assessed Quality.*

These two specifications are sufficiently alike that the following steps for selecting a sampling plan for reliability apply equally to either.

1 Decide on the reliability value which is to be demonstrated, and the corresponding higher value of reliability which the plan is expected to accept. Assign confidence levels to each. Thus suppose we have a batch of 200 parts, which we wish to demonstrate has a reliability of 93 per cent with 90 per cent confidence. We also wish to be reasonably certain that if the reliability exceeds 99 per cent the batch is unlikely to be rejected.

2 Convert this information to quality control terms. Thus 93 per cent reliability corresponds to $(1-0{\cdot}93) = 0{\cdot}07 = 7$ per cent defectives, to be accepted with a probability of $(1{\cdot}00-0{\cdot}90) = 0{\cdot}10$. A reliability of 0·99 corresponds to 1 per cent defectives.

3 For normal inspection, the tables corresponding to a batch size of 200 are given in table X–G of either specification. There the operating characteristics are drawn out, and we can see at once that our requirements correspond to an AQL of 0·40 per cent. The table below then shows that the sample size is to be 32 and that 0 failures can be permitted.

In the Table (p 238), the specifications provide for a range of inspection levels associated with different operating characteristics. If we decide to use these, we must remember that these specifications are concerned with making sure that the AQLs of the alternative plans are equivalent. They may differ appreciably in the value of reliability they will demonstrate.

19.9. recent developments in reliability demonstration

Because reliability demonstration tests are so expensive, various attempts have been made to economize without reducing the confidence which can be placed in the results. One of the most promising proposals is to make use of what we already know about the reliability of the items to be tested.

Ordinary reliability demonstration tests, which we discussed in section 19.8, assume that at the start of the test we know absolutely nothing about the reliability of the items submitted and that there is an equal chance that they could have any value from reliability 0 up to reliability 1·00. The test then proceeds to specify a suitable sampling plan to determine the reliability of the items with the required accuracy.

This ignores the fact that a lot of information is usually available at this stage. We probably already know the reliability of many of the units, etc. which make up the items to be tested and, by combining these together, we have a not too inaccurate prediction of the reliability of the complete item. For example, it may be that we can restate our test requirements as follows. 'We wish to demonstrate with 90 per cent confidence that this group of items has a reliability of at least 0·95. Our prediction says that their reliability should be 0·97, and we are 95 per cent confident that it is within the range 0·88 to 0·98.' Now if we are prepared to accept and use this additional information, we can now start with an assumed worst reliability of 0·88 instead of 0.

Some interesting work along these lines has been done by L. W. F. Lukis. The mathematics he uses is beyond the scope of this book, but is based on Bayes' theorem which we explained in sections 17.11 and 17.12. Readers who are interested may like to obtain a copy of his paper published in *The Quality Engineer* for September 1971, and entitled, 'Reliability assessment: simplified Bayesian methods'.

CHAPTER 20 MEASUREMENT OF RELIABILITY

20.1. the difference between reliability demonstration and reliability measurement

So far we have only considered tests to determine whether the reliability of a particular batch is likely to be greater than some preset figure. This as the Table on p 238 shows, if we test 15 and get 0 failures we are 90 per cent certain that the reliability of the batch is better than 0·85, and if we test 45, still with 0 failures, we are 90 per cent certain it is better than 0·95. However, a batch whose actual reliability is 1·00 is bound to give 0 failures, whatever the sample size. Hence all we know about the batch we have tested is that its true reliability is probably between the figure we have demonstrated and 1·00, but we cannot estimate its actual value unless we run the test long enough to get some failures. Therefore tests to estimate values of reliability are likely to be more expensive than demonstration tests.

Although we could run a test to measure reliability itself, it is more likely that we would estimate one of the following parameters, which we could then convert to reliability if required:

1 The rate of degradation or deterioration.
2 The mean time between random failures (θ).
3 The mean time between wear out failures (M), and the variability about this mean time (σ).
4 The parameters of the Weibull distribution.

We shall now consider in detail how to conduct each of these tests.

20.2. the rate of degradation

Many parts operate satisfactorily for a time, possibly a very long time, and then without warning they fail suddenly. Sometimes, however, we can observe a gradual deterioration or degradation and so see a failure coming. For example, two mechanical parts which slide upon each other might be examined at intervals for signs of wear. Hence the rate at which they were deteriorating could be estimated long before final failure occurred.

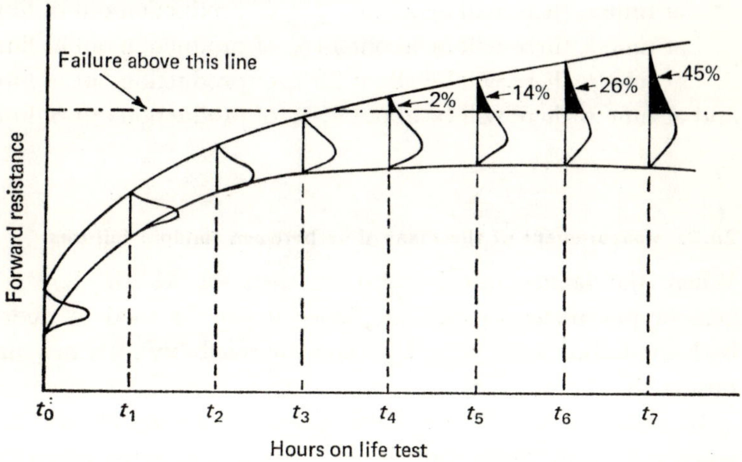

FIG 20.2(a). Degradation of selenium rectifiers on life test

In Fig. 18.5(a) we drew a typical deterioration curve for the forward resistance of selenium rectifiers. Suppose now that we have a sample of rectifiers which we wish to life test. Even at the start of the test their forward resistances will not all be identical, and if there are enough of them we shall be able to plot a frequency distribution curve. For technical reasons this is unlikely to be quite Normal, and will probably be skewed somewhat as Fig. 3.5(d). This curve is shown at t_0 in Fig. 20.2(a). It should really stand upright with its base aligned to time t_0, but as this is clearly impossible we have drawn it flat.

As the test proceeds each rectifier will not only deteriorate, but they will also tend to vary more one to another. Hence the bases of successive distribution curves become longer, and their peaks get lower. If the curves are well defined, we may be able to predict by about time t_2 roughly what will happen in the rest of the test. This clearly saves a lot of both time and expensive testing where routine work is involved. No doubt from time to time we shall let a test run its full length, in order to confirm that our predictions remain correct.

If the maximum permissible forward resistance after deterioration is given by the line shown, then we can predict that

at time t_4 there will be about 2% of production out of limit,
at time t_5 there will be about 14% of production out of limit,
at time t_6 there will be about 26% of production out of limit,
and at time t_7 there will be about 45% of production out of limit.

20.3. measurement of the mean time between random failures

Where the failure rate is constant, then the MTBF is a convenient parameter to estimate, since it can be used to deduce both the failure rate ($\lambda = 1/\theta$) and the reliability after any time t using $[R = \exp - t/\theta]$.

In principle we put a suitable sized sample on test and note when each fails. Some care is necessary however to ensure that the failure rate remains constant throughout the period of observation, and this means that the test must be confined to the constant failure rate period of the bath tub curve. Thus:

1 The items to be tested must first be operated under test conditions, until they have completed their early failure period OA, and have entered period AB of Fig. 2.1(a).
2 The test must be completed before the wear out period starts at B.
3 The items must collectively accumulate the required number of test hours, but as explained in section 19.5, we can arrange these to suit ourselves. Thus if we require 100,000 hours of test time, we can test 100 items for 1000 hours, or 1000 items for100 hours, etc., as we choose.
4 The items tested need not be new, and in general cannot be, because of the need to avoid the early failure period.
5 Often it is convenient to test a fairly large number of items for a relatively short time. There is no necessity to continue

the test until all have failed, and to do so may run the risk of incurring wear out failures.

6 Items which fail can be replaced with good ones so as to accumulate the required number of hours more quickly.

20.4. procedure for estimating the MTBF

Our test procedure should be broadly as follows.

1 DECIDE PRECISELY

a The conditions of test. This will include stresses such as the mechanical load, applied voltage, etc., and the test environment expressed in terms of the ambient temperature and humidity, etc.

b What constitutes a failure. Otherwise there may be arguments later on as to whether a particular item has failed or not.

2 ALLOCATE TEST RIG SPACE This was discussed in section 18.1.

3 SELECT A RANDOM SAMPLE This means that every item from which the sample is to be drawn must have an equal chance of being included in it.

4 CARRY OUT THE TEST We now put the items concerned on test, making sure that all the conditions are controlled within the limits we have set. In principle we note exactly when each item fails, and replace it with a good one. In practice this may not be easy, because the test will presumably run 24 hours a day for 7 days a week. Indeed this normally must be so, because any attempt to switch off in the evening and restart next morning would almost certainly lead to increased failure rates. However, it may not be convenient to have someone there all the time on the alert to note the exact moment of each failure. If the test justifies it, it may be possible to install an automatic device to do this, but in many cases it will be sufficient to inspect the test rig at a preset time, say twice a day. Thus suppose we inspect at 7.00 and 19.00 hours each day. If we detect a failure at 7.00 hours,

we shall assume that it failed midway between 19.00 hours the previous evening and now, and assign its failure to 1.00 hours in the morning. Similarly failures discovered at 19.00 hours will be assigned to 13.00 hours. Since the failure rate is assumed to be constant, the cumulative error from doing this will tend to be zero and, as most tests run over several thousands of hours, the overall error is likely to be small.

5 WORK OUT THE RESULTS When sufficient failures have occurred, we stop the test and work out the results as discussed in the next section.

6 EXAMINE THE ITEMS AFTER TEST Since reliability tests are expensive, we shall want to extract as much information as possible from them. In many cases it is profitable to carry out a post mortem to determine why those items which failed did so, and equally to try to understand why those which survived did not succumb like the others.

20.5. evaluation of the results of an MTBF test

Tests are usually stopped or *truncated* when sufficient data has been gathered. We then add up the time that every item has been on test, regardless of whether it has failed or not, and divide by the number of failures.

Thus consider a simple example where 10 items are put on test. Three fail after 200, 1100 and 1900 hours from the start, respectively, and are not replaced. The remaining seven survive until the test is truncated at 2500 hours. Then:

$$\text{Total hours on test} = 200+1100+1900+(7\times 2500) = 20{,}700 \text{ hours}$$

$$\text{Total number of failures} = 3$$

$$\text{Observed MTBF} = \hat{\theta} = \frac{20{,}700}{3} = 6900 \text{ hours}$$

We can express this as a formula. Thus suppose:

a We test n items.

b The test is truncated at time t, which is after the rth item has failed, but before the failure of the $(r+1)$th failure.

c The times at which each item failed are denoted by t_1, t_2, t_3, etc., up to t_r. Then :

$$\hat{\theta} = \frac{t_1+t_2+t_3+\cdots+t_r+(n-r)t}{r} \qquad (20.5a)$$

If each item which fails is immediately replaced with a good one, so that there are always n on test, then :

$$\hat{\theta} = \frac{nt}{r} \qquad (20.5b)$$

If we could accurately test every item of a particular type which ever existed then we would know exactly what its MTBF was. In practice this is impossible, and we must test what is often a very small sample and hope that this will give us a reasonable idea of the true MTBF. We must therefore consider how accurate or otherwise any estimate which we have made is likely to be. Since we are concerned with an exponential distribution, we shall need to use the chi squared distribution, which is usually written χ^2. There is no need for us to consider its theory, because we can set out the way it is to be used in three simple steps as follows.

1 To calculate the confidence limits we need to know :

The number of failures which were observed $= r = 3$
The observed MTBF $= \hat{\theta} = 6900$

The confidence level required. Suppose we decide that this is to be 90 per cent.

2 Now a confidence level of 90 per cent means that we are prepared to take a 10 per cent risk that the true value of the MTBF is outside the confidence limits we give. If we divide this equally, it means that there will be a 5 per cent risk of being outside each limit, and hence that our 90 per cent confidence is to spread from 5 to 95 per cent. These values are designated by $\alpha = 0{\cdot}05$ and $\alpha = 0{\cdot}95$.

3 We now consult tables which give the percentage points of the chi squared distribution, such as *Statistical Tables for Science, Engineering and Management* by J. Murdoch and J. A. Barnes which is published by Macmillan.

UPPER CONFIDENCE LIMIT

We insert in the table the value of $\alpha = 0{\cdot}95$ and the number of degrees of freedom ν, which for our purposes is $\nu = 2r = 2 \times 3 = 6$.

Hence inserting $\alpha = 0{\cdot}95$ and $\nu = 6$, we read that $\chi_u^2 = 1{\cdot}635$. To calculate the upper confidence limit, we use:

$$\text{Upper confidence limit} = \frac{2r\hat{\theta}}{\chi_u^2} = \frac{2\times3\times6900}{1{\cdot}635} = 25{,}300$$

LOWER CONFIDENCE LIMIT

The lower confidence limit has a similar calculation. As we said $\alpha = 0{\cdot}05$, and if the test is terminated at failure number r, we again use $\nu = 2r = 6$. If as in this case the test is continued until sometime between failure r and $(r+1)$, then strictly speaking we should use $\nu = (2r+2)$. Hence $\nu = (2\times3)+2 = 8$, and inserting $\alpha = 0{\cdot}05$ and $\nu = 8$ into the χ^2 tables, we read that $\chi_L^2 = 15{\cdot}507$. Therefore:

$$\text{Lower confidence limit} = \frac{2r\theta}{\chi_L^2} = \frac{2\times3\times6900}{15{\cdot}507} = 2670$$

Thus we conclude that the mean time between failures, as far as we can judge from this small sample, has a best estimate of 6900 hours, and we are 90 per cent confident that the true value will not exceed 25,300 hours nor be less than 2670 hours.

Clearly this has not been a very precise test, and this is because we observed so few failures. If we had carried out the test on a larger scale, then the confidence limits would be nearer together. They would also be nearer if we reduced our confidence level to say 60 per cent.

20.6. an example of the calculation of the MTBF

1 TEST OBSERVATIONS

Suppose we put 100 items on life test. We note when each fails, but decide not to replace them. Failures are observed at the following times, counting the start of the test as time $t = 0$.

3	29	128	296	429	653	851	1027 hours
13	34	161	353	473	670	880	
19	39	167	373	514	684	901	
21	65	200	392	558	711	953	
22	84	256	422	635	773	967	

The test was truncated after 1027 hours, at the moment when the 36th failure occurred.

2 ELIMINATING EARLY FAILURES

Almost at a glance, we are bound to be suspicious about the number of failures which occur very early in the test. Thus overall some 36 failures occur in 1027 hours, which is an average of 1 failure every 28 hours. In the first 84 hours, however, there are 10 failures at an average of 1 every 8·4 hours. It is true that the number of items on test drops slightly as the test proceeds, but not to that extent. It therefore seems likely that the first 80 or so hours of the test contain some early or burn-in failures, represented by the side of the bath tub curve marked OA in Fig. 2.1(a). To be safe therefore we must exclude these early failures from our MTBF calculations, although they can of course be used to give an indication of both the density and the duration of the early failure period. Thus the above results indicate that about 10 per cent of the items may fail during the early failure period, and that to remove them will require a burn-in period of 100 hours or a little more.

Suppose we decide to exclude all failures up to hour 84. This gives us:

10 early failures which we exclude,
26 failures during the relevant part of the test, and
64 items which survived the test.

We thus have useful results on $26+64 = 90$ items.

3 CHECKING FOR A CONSTANT FAILURE RATE

We may wish to check that the failure rate is in fact approximately constant, since if this is not so any estimated MTBF will have little meaning. In section 3.3 we showed that where items which fail are immediately replaced, so that the number on test is always constant, we merely have to plot a graph as Fig. 3.3(a) showing the total number of items which have failed so far against the time since the test started. If the failure rate is constant, this will yield a straight line.

We can easily adapt this for the case where the items are not replaced, by plotting horizontally the total number of test hours for all the items added together. Thus, as far as we are concerned, the test proper started at time 84 hours with 90 active items, and all of these lasted until the first valid failure at time 128 hours. Therefore: Total test hours from 84 to 128 hours $= 90 \times (128-84) = 3960$. Now there are only 89 active items. Thus: Total test hours from 128 to 161 hours $= 89 \times (161-128) = 2937$, giving $3960+2937 = 6897$ hours since the test started. If we continue this calculation, we can then plot Fig. 20.6(a) as

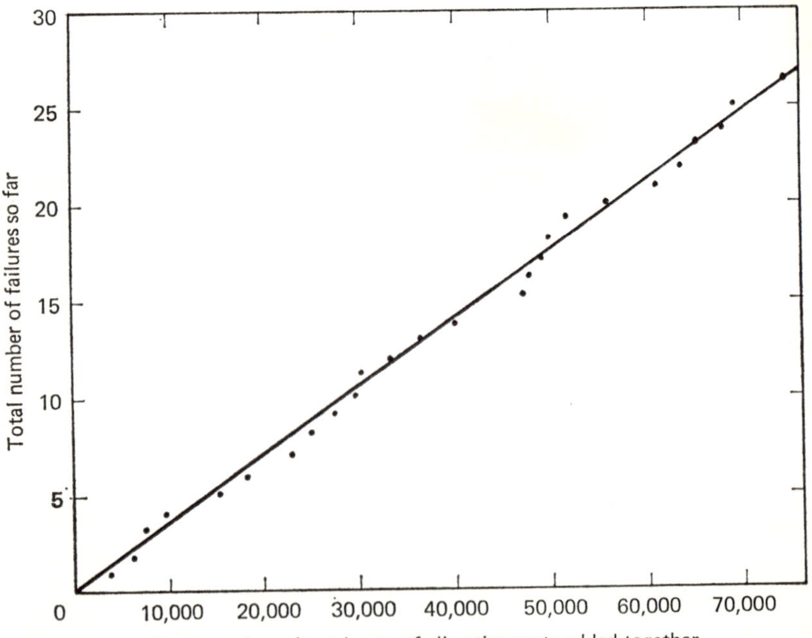

FIG 20.6(a). Example of a check for constant failure rate

shown, and the straight line confirms that the failure rate is indeed constant over the valid portion of the test.

4 CALCULATE THE MTBF

To estimate the MTBF we need to know the total number of hours that all the items have collectively been on test. If we have done the calculation in sub-section 3 above, we shall already have the total as 72,595 hours. If not we can calculate it by applying equation (20.5a) to the data. Thus:

a We add up all the times of failure, as given in sub-section 1, omitting only those up to and including hour 84, which we decided to exclude.
This gives — 14,427 hours

b We add on 64×1027 for the 64 which survived — 65,728 hours

The total is now — 80,155

c We subtract 84×90, because we decided not to count the first 84 hours of the test, which all 90 of the valid items survived. Therefore:
Subtract — 7,560
Leaving — 72,595 hours

d Our estimate of the MTBF is then

$$\text{MTBF} = \hat{\theta} = \frac{72{,}595}{26} = 2792 = 2800 \text{ hours, say}$$

5 CALCULATION OF THE PROBABLE ACCURACY

Suppose we again decide upon a confidence level of 90 per cent.

Then we have:

$$\alpha = 0{\cdot}05 \text{ and } 0{\cdot}95$$
$$r = 26$$
$$\hat{\theta} = 2800 \text{ hours}$$

Using the χ^2 distribution, exactly as described in section 20.5, this gives us that the two confidence limits are at approximately 2100 and 4000 hours.

6 CALCULATION OF THE FAILURE RATE

Since the MTBF is 2800 hours, therefore the failure rate is given by:

$$\hat{\lambda} = \frac{1}{\hat{\theta}} = \frac{1}{2800} = 0{\cdot}000{,}357 = 0{\cdot}0004 \text{ failures per hour, say.}$$

20.7. an alternative calculation for the MTBF

The MTBF can be calculated more easily but less satisfactorily as follows. We know that at the end of the test 64 items had survived out of the 90 which we decided to count, therefore

$$\text{Observed reliability} = R(t) = \frac{64}{90} = 0{\cdot}71$$

Now the duration of the valid part of the test was $= (1027-84) = 943$ hours

If we assume that the failure rate is constant, we can use equation (3.2b).

$$R(t) = \exp(-\lambda t) = \exp\left(-\frac{t}{\theta}\right) = \exp\left(-\frac{943}{\theta}\right) = 0{\cdot}71$$

From exponential tables

$$0{\cdot}341 = \frac{943}{\theta}$$

Therefore:

$$\hat{\theta} = 2766 \text{ hours.}$$

This figure of 2766 hours is a little lower than that of 2792 which we deduced by the first method. The agreement is quite good, but as both values were deduced from the same data we might at first sight expect them to be identical. In fact, however, they are not based on the same assumptions. The figure of 2792 hours obtained from the longer method represents the actual MTBF which was observed, with no assumption that the failure rate was constant. The second method disregards much of the data, and merely uses the reliability as observed after 943 hours. It then makes up for the deficiency in data by assuming that the failure rate is constant.

20.8. tests to estimate the mean time to wear out failure

A test to estimate the mean time to wear out failure is basically similar to the steps set out in section 20.4, but with the following important differences:

1 As can be seen from Fig. 2.4(a), wear out failures for a particular part tend to occur around a set interval OM after a new part is put on test. Therefore for this test the parts used must be new. Compare this with the mean time to random failure where parts did not have to be new, so long as the test period was confined to the flat constant failure rate part of the bath tub curve.

2 Since we must start with new parts, we are likely to incur both early failures and random failures (OA and AB respectively of Fig. 2.4a) on our way to the true wear out failures. Where possible, we shall try to eliminate these.

3 We must continue the test until all or almost all of the parts have failed from wear out, and therefore we must be prepared for a long test.

4 The number tested will probably be less than it would be for a corresponding random failure test. It will be based on the number of failures we need to observe. We must however allow for those which will be eliminated as early or random failures. If the useful working life of the parts is long (i.e. OB of Fig. 2.4a), an appreciable fraction may be lost in this way. There is also the risk that the wear out failures, when we finally reach them, will not form any recognizable wear out pattern such as a normal frequency distribution curve. This might be because too many parts have been lost on the way or it may be that wear out failures are due to more than one cause, and that each is trying to form its own frequency distribution curve.

5 Parts which fail should not be replaced. If we lose so many from early and random failures that we are forced to put some more on test, we must for these start counting the time all over

again even though by then the rest of the parts may have run for several thousand hours.

20.9. calculation of the mean time to wear out failure

There are two quantities which we shall need to deduce from our test results:

1 The mean time to wear out failure $\hat{M}$.
2 The standard deviation about that mean time $\hat{\sigma}$.

The steps in the calculation are as follows:

1 Eliminate failures which appear to be either early or random. This is not always easy, but the following points may help:

a Failures which are well clear in time from the cluster of failures at wear out are probably nothing to do with wear out, or their failure is due to a different cause from the rest.
b Sometimes wear out failures show obvious signs of wear or other deterioration, and it is possible to sort the data after the failures have been examined.
c Random failures do of course continue until the test is complete although they reduce, particularly during the wear out period, as the number on test reduces. Fortunately in many cases their inclusion in the wear out calculations does not make a lot of difference to the results.

2 We calculate the mean time to wear out failure, and assign confidence limits to it at a suitable confidence level. Thus:

$$\text{Mean time to wear out} = \hat{M} = \frac{\text{Total time on test of all parts which failed from wear out}}{\text{Total number of parts which failed from wear out}}$$

If we let $t_1, t_2, \ldots, t_n$ be the times to each wear out failure, then:

$$\hat{M} = \frac{t_1 + t_2 + t_3 + \cdots + t_n}{n} \qquad (20.9a)$$

Since we are dealing with a Normal distribution, our confidence limits must be determined using the Student's t distribution. We shall explain how to do this when we work out an example in section 20.10.

3 Next we calculate the standard deviation exactly as we discussed in section 3.6 using equation (3.6b), which becomes:

$$\hat{\sigma} = \sqrt{\frac{\Sigma(t_f - \hat{M})^2}{n-1}} \qquad (20.9b)$$

The example in the next section shows in detail how this is done.

4 As we explained in section 3.6, the mean time to wear out failure M gives us the time at which approximately half the parts will have failed. We are usually more interested in the time at which wear out failure starts, because this is the latest time at which we can replace our parts if we wish to avoid wear out failures. This is given by $(M-3\sigma)$.

20.10. an example of the calculation of the mean time to wear out failure

Suppose we put 30 parts on accelerated life test, and they fail at the following times from the start of the test:

19	187	472	714	802	831
40	234	490	762	812	843
51	301	533	775	812	847
97	317	608	790	814	870
118	394	667	800	829	958

We can see at once that we have got some sort of a wear out failure pattern towards the end of the test, but it is not easy to visualize exactly what it is. It often helps to draw a simple graph, as shown in Fig. 20.10(a), where each dot represents a failure. We can now see that the first 3 failures might well have belonged to the early failure period, and that these are followed by some 13 failures which are probably random in origin. Thus we shall eliminate the first 16 failures, up to and including that at 714 hours, as being unlikely to be due to wear out. The failures from

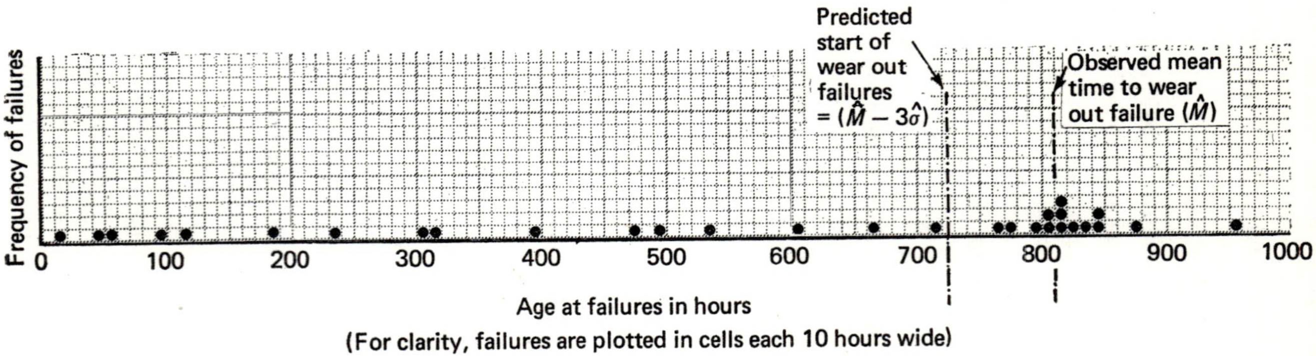

FIG 20.10(a). Plot of the test results in the example

762 to 870 hours appear to be part of the wear out pattern, although they may well contain 1 or 2 random failures. If we can identify these we shall eliminate them, but probably their inclusion will not make a lot of difference and we should certainly not discard any result merely because we have a 'hunch' about it. There is one very long life of 958 hours which should not be included. It is obviously not typical of the rest, and its inclusion might give the impression that the mean time to wear out failure is longer than it really is.

We therefore average the failure times from 762 hours to 870 hours, and this gives:

Total hours parts were on test = 10,587
Number of failures observed = 13
Observed mean time to wear out failure $= \hat{M} = \dfrac{10{,}587}{13} = 814$ hours

Next we estimate the standard deviation

	$(t_f - \hat{M})$	$(t_f - \hat{M})^2$
762−814 =	−52	2704
775−814 =	−39	1521
790−814 =	−24	576
800−814 =	−14	196
802−814 =	−12	144
812−814 =	−2	4
812−814 =	−2	4
814−814 =	0	0
829−814 =	15	225
831−814 =	17	289
843−814 =	29	841
847−814 =	33	1089
870−814 =	56	3136

$$\sum(t_f - \hat{M})^2 = 10{,}729$$

Observed standard deviation $\hat{\sigma} = \sqrt{\frac{\Sigma(t_f - \hat{M})^2}{n-1}} = \sqrt{\frac{10{,}729}{13-1}}$

$= 30$ hours approx.

To estimate the time at which wear out is likely to start, we use equation (3.6c).

$$\begin{aligned}\text{Start of wear out failures} &= (\hat{M} - 3\sigma) \text{ approx.}\\ &= 814 - (3 \times 30)\\ &= 724 \text{ hours}\end{aligned}$$

We must however consider how accurate our results are. Since the mean time to wear out is assumed to relate to a Normal frequency distribution, we must use a technique called Student's t as follows. Let:

n = Number of wear out failures observed

t = The value we require from the percentage points of the t distribution, which we shall take from the same tables as we used in section 20.5. As before, if the confidence level is to be 90 per cent we must insert 0·05 in the tables. In this case, however, the value of ν is $(n-1)$. Therefore $\nu = (13-1) = 12$.

The relation we now use is:

$$\begin{aligned}\text{Upper and lower confidence limits} &= \hat{M} \pm t \times \frac{\hat{\sigma}}{\sqrt{n}}\\ &= 814 \pm \frac{1{\cdot}782 \times 30}{\sqrt{13}}\\ &= 814 \pm 15\\ &= 799 \text{ to } 829 \text{ hours}\end{aligned}$$

20.11. an example of the estimation of the parameters of the Weibull distribution

Suppose we put 24 items on life test, and by noting when each fails we obtain the following data:

1450	2290	3000	3530	3900	4870
1800	2540	3000	3610	4080	5400
1970	2540	3280	3680	4500	5650
2080	2690	3470	3780	4670	6200

We can tell at a glance that the failure rate cannot possibly be constant, and this is confirmed by Fig. 20.11(a) where failures are plotted against a linear time scale. Thus at the start there is a gap of 1450 hours before any failures occur. Between 3000 and 4000 hours there are a lot of failures and then they thin out and cease altogether after 6200 hours, when the last fails. The distribution is not Normal either, for it is clearly skewed towards its upper end. It is not essential to plot Fig. 20.11(a) in practice, but it is useful for the purposes of our explanation.

Where neither the exponential nor Normal distributions fit, we naturally consider the possibilities of the more flexible Weibull distribution which we discussed in sections 3.11 to 3.13. In order to use Weibull probability paper, we must plot the age at failure against the cumulative percentage of failures. In working the latter out, we introduce a correction factor for the smallness of our sample. This involves adding 1 to the total number of failures observed, so that in our case we shall base our percentages on $24+1 = 25$. This gives:

Failure	*Cumulative percentage failures*
1st	1 in 25 = 4%
2nd	2 in 25 = 8%
3rd	3 in 25 = 12% and so on

We can thus plot our Weibull curve using the data given in columns B and C of Table 20.11(a) below.

We obviously hope that this will yield a straight line, but in this case our luck is out and the points contain a distinct curve. The straight line AB in Fig. 20.11(b), which coincides roughly with the points at the top end of our plot, illustrates our difficulty.

Now a Weibull plot is only straight if the location or minimum

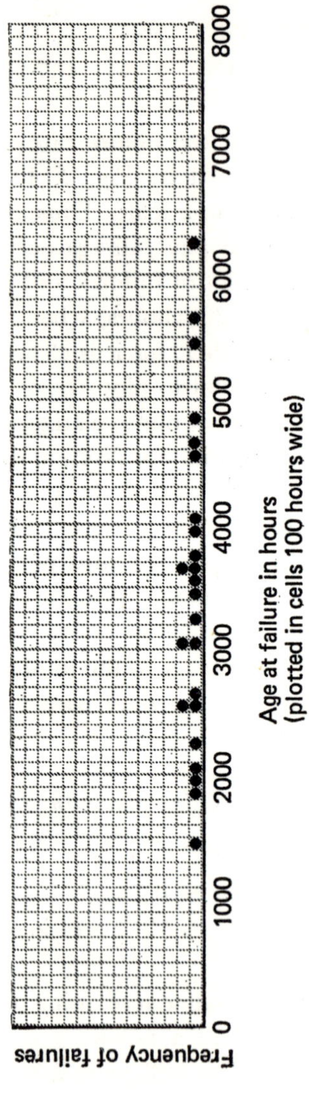

FIG 20.11(a). Plot of test results on ordinary equal division graph paper

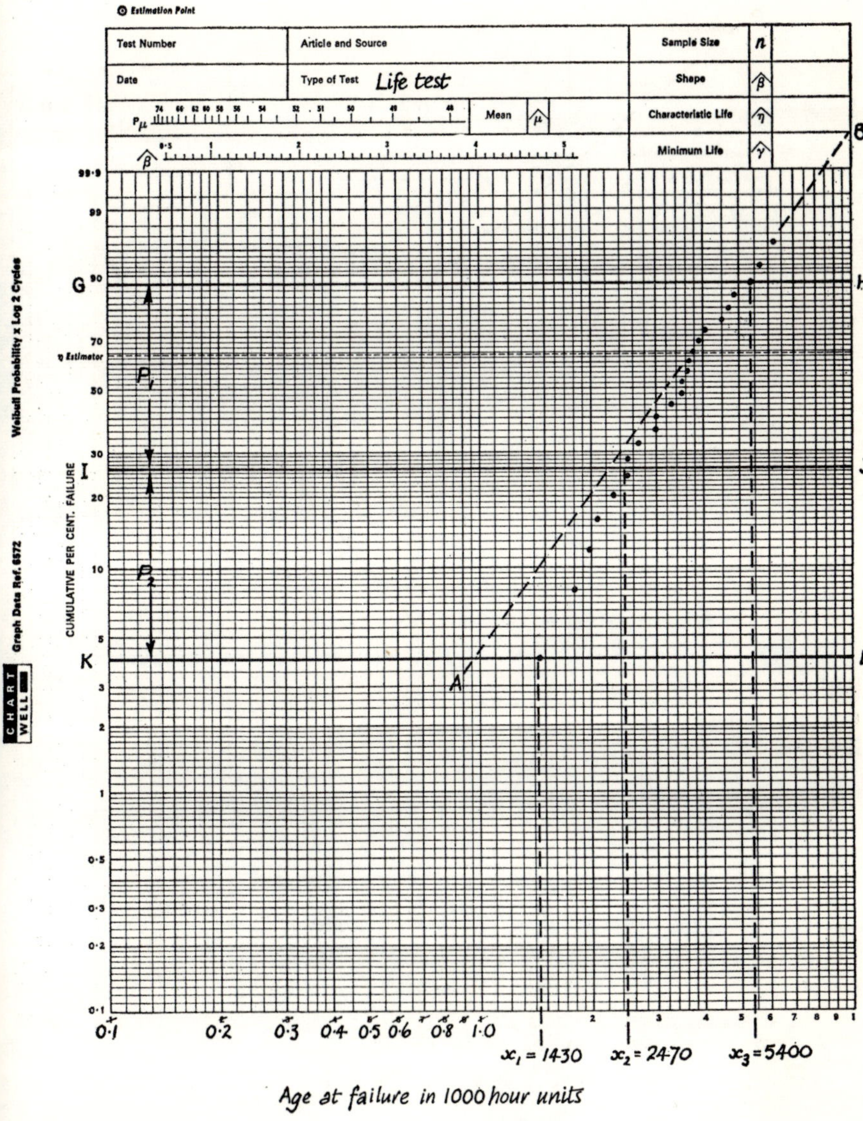

FIG 20.11(b). Weibull plot of data as observed

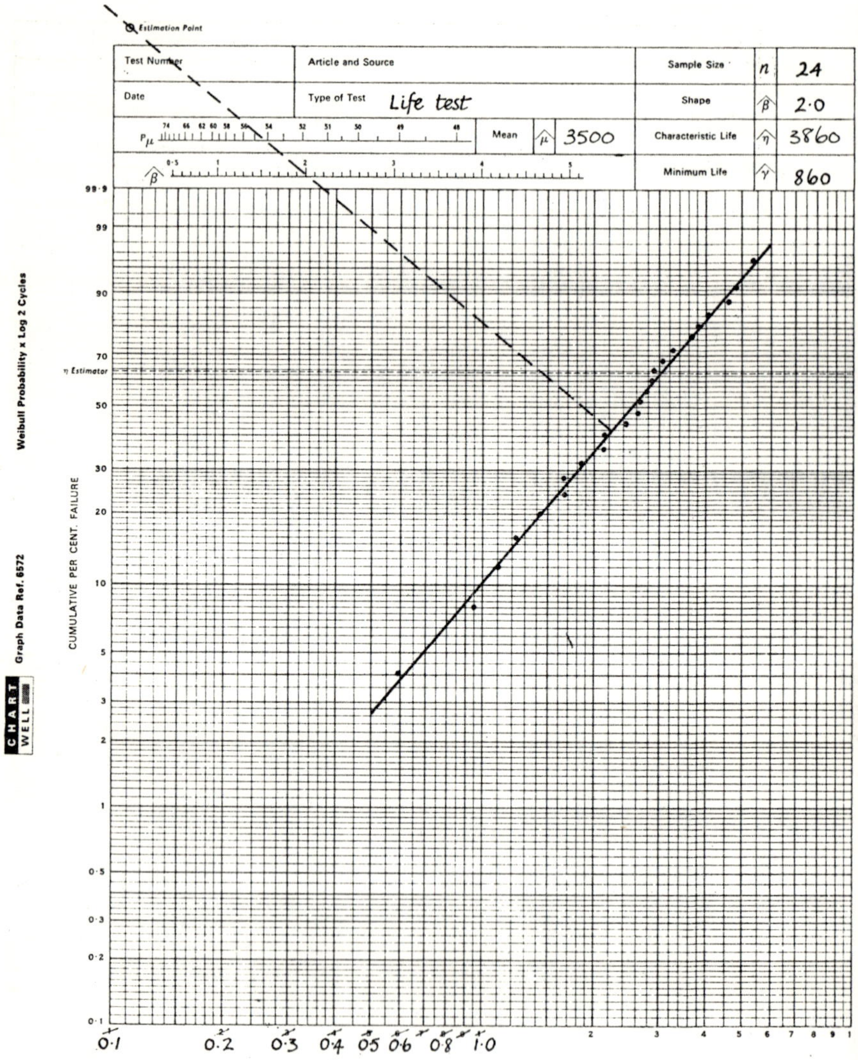

FIG 20.11(c). Weibull plot after data has been adjusted so that $\gamma = 0$

life parameter $\gamma = 0$. As this is probably not so in this example, we calculate $\hat{\gamma}$ as follows:

1 On the Weibull plot in Fig. 20.11(b), we draw any 3 horizontal lines GH, IJ and KL, so that they are equally spaced in millimetres (not equal in terms of the Weibull scale). Thus $P_1 = P_2$ in millimetres.

2 Where these lines cut the Weibull curve, we drop verticals on to the age at failure scale as shown, and read the values of x_1, x_2 and x_3. Thus in our example we get:

$$x_1 = 1430$$
$$x_2 = 2470$$
$$x_3 = 5400$$

We now use the following expression to estimate γ.

$$\hat{\gamma} = \frac{(x_1 x_3 - x^2{}_2)}{(x_1 - 2x_2 + x_3)} = \frac{(1430 \times 5400) - (2470)^2}{(1430 - 4940 + 5400)}$$
$$= 860 \text{ hours approx.}$$

Hence we must deduct 860 hours from each value in column C in the table shown overleaf to get the figures in column D.

We now plot a new Weibull curve, using the data in column D, and this is shown in Fig. 20.11(c). This time a straight line results and we can estimate the Weibull parameters as explained in section 3.13. Thus:

The η estimator line gives $\hat{\eta} = (3{\cdot}0 \times 1000) + 860 = 3860$ hours.

A perpendicular from the estimation point onto the Weibull line gives

$$\hat{\beta} = 2{\cdot}0$$

We already have that $\hat{\gamma} = 860$ hours

Hence we now have estimates of all the Weibull parameters.

A *Failure number*	*B* *Cumulative failures (%)*	*C* *Actual age at failure*	*D* *Age at failure less 860 hours*
1	4	1450	590
2	8	1800	940
3	12	1970	1110
4	16	2080	1220
5	20	2290	1430
6	24	2540	1680
7	28	2540	1680
8	32	2690	1830
9	36	3000	2140
10	40	3000	2140
11	44	3280	2420
12	48	3470	2610
13	52	3530	2670
14	56	3610	2750
15	60	3680	2820
16	64	3780	2920
17	68	3900	3040
18	72	4080	3220
19	76	4500	3640
20	80	4670	3810
21	84	4870	4010
22	88	5400	4540
23	92	5650	4790
24	96	6200	5340

Table 20.11(a)

PART FIVE **RELIABILITY PREDICTION AND MANAGEMENT**

CHAPTER 21 RELIABILITY PREDICTION AND CONTROL

21.1. the need for reliability control

Having now discussed all the main aspects of reliability, we must consider how we can unite them so that when a system is finally in operation it really does achieve the reliability its designers intended. This will only be possible if reliability is closely controlled at every stage, from the original conception of the project through research, design, development and manufacture up to and including the installation and operation of the system by the ultimate user.

21.2. feasibility studies

Where a customer requests a specialized piece of equipment, it should be possible to agree the customer specification with him. In the more general case, ideas for new projects may come from a variety of sources, for instance :

Research and development reports.
Reports from salesmen, service engineers or market surveys.
Requests and suggestions from customers.
The purchase of patents.
Suggestions from production personnel.

Before we embark on a new project of any size, we shall want to satisfy ourselves that it has a good chance of achieving its intended performance and reliability within the time available and at an acceptable minimum cost. A quick decision is seldom possible. Promising ideas must first be submitted to an investigation to determine their feasibility and to estimate their probable cost. A commercial organization will want to assess profitability. Sometimes it is desirable to subject alternative possibilities to a preliminary investigation, to decide whether a full scale feasibility study is likely to be worthwhile.

We must try to foresee the situation that will arise if we fall down on any of our objectives. If the predicted profitability is so finely balanced that a small rise in the development costs would make the whole project unprofitable, it may be better to abandon it altogether.

Preliminary analysis and assessment will usually be interspersed with research and development tests and the collection of other information on which to base a decision. Even so, approval at this stage may be no more than a go ahead for a preliminary design, after which the project will be reviewed again.

21.3. the preliminary design

The importance of the preliminary design is sometimes overlooked. Decisions made at this stage can affect the expenditure of thousands of pounds at a later date. Subsequent modification to improve performance or reliability is often both difficult and expensive, and experience shows that those concerned with reliability are seldom very satisfactory. Thus it is important to ensure as far as possible that everything is right as early on as possible.

Often there are several possible design alternatives from which the most promising must be selected. Our approach to this will probably be along the lines discussed in section 7.2. We shall prepare a block or unit diagram for each alternative. Where a block or unit has been made before, its performance and reliability should already be known. Where this information is not available, a target performance and reliability must be assigned to it. Since the blocks will normally be in series in the reliability sense, we can then utilize equation (7.2a). Thus:

$$\text{System reliability} = R_s = R_1 \times R_2 \times R_3 \times \cdots \times R_j$$

where R_1, R_2, etc. represent the reliabilities of each block in turn. Alternatively if we are working in failure rates, we have:

$$\text{System failure rate} = \lambda_s = \lambda_1 + \lambda_2 + \lambda_3 + \cdots + \lambda_j \quad (21.3a)$$

If the target value for the system reliability (or failure rate) has been set, then we start from this and allocate reliabilities to each block, allowing the lowest figures where difficulties are likely to be encountered. If we have no overall target, then we allocate probable reliabilities to each block and calculate the predicted system reliability using equation (7.2a).

Such predictions may not be very accurate. It depends how far they are based on fact and how far on guesswork. Fortunately it is often possible to select the best of several alternatives, even though the actual value of its reliability may still be in doubt.

Where, as in some military applications, it is essential to achieve the reliability demanded even if the resulting cost is high, then it is usually best to predict a little pessimistically. If the reliability finally turns out to be higher than predicted, it is all to the good. In the more general case where performance, reliability and cost are more nearly equal in importance a midway approach will probably be better. Achieving a higher reliability than was demanded, at a higher cost than was forecast, may make the product unsaleable. In such cases it may be prudent to keep to the predicted cost, and provide the best performance and reliability that it will achieve. It is then comforting to remember that the actual reliability in customers' hands sometimes exceeds the prediction. This may happen where many customers do not use the equipment concerned at its full rating or under the most arduous of conditions.

21.4. producing a detailed design from which a prototype can be prepared

At a relatively early stage, therefore, a decision is usually made about which of the various design alternatives will be adopted. This is a convenient moment to review all the predictions concerning reliability, performance, costs and timing. If everything is still favourable a decision is made to go ahead.

At regular intervals the reliability manager will meet representatives from all the interested departments, such as design, development, production, test and finance, and review progress with them. He will have to ensure that resources are concentrated on sections of the design where it appears that adequate reliability may not be achieved. In principle there are a number of things which may be done:

1 If he finds that some blocks can achieve a higher reliability than expected, he may be able to reallocate the reliabilities so as to permit lower reliabilities in the troublesome areas.
2 Further studies of the design may be necessary, probably accompanied by tests and development work.
3 Redundancy may be used to improve the reliability of the critical sections. This might involve the duplication of a complete block, but it is more likely that only those units or elements which have the worst failure rates will be duplicated.

It is often useful to make three reliability predictions for each block:

1 The reliability expected from the existing design in its present form.
2 The reliability anticipated from improvements which can be immediately foreseen and which are broadly within the present 'state of the art'.
3 The ultimate reliability which could be achieved if all systematic failures were eliminated. (See section 2.3).

Probably this will never quite be reached, but at least it indicates the limitations of the design. If we need a reliability which is better than this ultimate prediction, then we shall probably have to consider radical design improvements.

21.5. prototype tests

Tests to try out new principles or to ascertain failure rates will no doubt go on in parallel with design, as we explained in section 12.1. As sections of the designs are released, however, a formal and independent series of prototype tests will be possible. Units or blocks will probably be tested individually at first, and then when it appears that each is satisfactory they will be assembled together for a test on the complete system.

Activating each unit of the equipment will be done broadly on the lines which we discussed in section 15.4 for the commissioning of a new piece of equipment, but with more precautions and more attention to detail. A new piece of equipment which has been sold to a customer is usually known to work if correctly built and operated. With a prototype we have no such guarantee. It may contain some design errors which make correct functioning impossible. We therefore want to start it up in such a way that if there is anything wrong we both avoid damage to the system and trace the cause of failure.

In principle, prototype testing offers a moment of truth. Now at last we really know whether our earlier predictions were correct. Even so it is easy to be misled. Thus :

1 If the prototype was very carefully made by craftsmen specially selected for their skill and has been subsequently operated by highly trained technical staff, it may perform far better than we can hope from ordinary production models. Results are probably much more akin to the inherent reliability of the design which we discussed in section 12.1.

2 Sometimes however the prototype contains various 'make-do' components, which will be much improved later when produced from specially designed tools. In this case it may be that production models will actually be better than the prototype. We must be careful however not to allow every shortcoming to be glossed over with the statement that, 'It will be all right when we get the proper tooling.'

In general the designers should not do their own prototype testing. It is better for a neutral and independent team to do this. Thus prototype testing might conveniently be the responsibility of the reliability manager or the quality manager and their staff. The designer must however be completely answerable for the performance and reliability of his own designs, and if tests reveal deficiencies it is the designer's job to put them right.

Sometimes it is convenient to divide defects initially into three groups:

1 *Design errors* where the prototype is to drawing, but still does not work satisfactorily.
2 *Manufacturing errors* where the prototype has been incorrectly made.
3 *Part faults* where purchased parts are defective.

We must of course be careful to carry out our prototype tests under the correct environmental and other use conditions, otherwise the results may mean very little.

21.6. manufacture and production testing

We must continue reliability prediction and allocation right through manufacture and production testing. At every stage we compare the reliability which has been achieved with the prediction and, if the former falls short, we take corrective action. Indeed this must go on right through to the site commissioning tests and subsequent customer usage. If we relax at any stage, we may find that actual reliability is below what is required.

When the first feasibility study is done at the start of the project, our prediction will probably be somewhat inaccurate. At every subsequent stage we shall be a little more certain of what the final reliability will be. At last in the customer's hands we have the ultimate truth. If liaison with our customers is good, we shall be able to find out precisely what we have achieved and whether there are any shortcomings.

CHAPTER 22 RELIABILITY MANAGEMENT

22.1. reliability objectives

In this final chapter we must consider the organization and control which we shall need in order to reach our reliability objectives. This will vary with the type of work we do, and in particular it will depend upon whether we are concerned with system manufacture or utilization. Nevertheless our objective is likely to be to achieve a satisfactory standard of reliability under the required operating conditions:

1 As cheaply as possible.
2 At the time required.
3 With due regard for other relevant factors such as performance, appearance, weight, size, maintainability, etc.

22.2. the people concerned with reliability

To achieve this objective we shall need to put into practice everything in this book which is appropriate to our particular circumstances, and adapt it as necessary. Fig. 22.2(a) therefore summarizes the whole of reliability, and includes references to the various chapters in this book where further information can be found. We can see at once that a very large range of people are involved. At the start of a project, we have the market specification, emphasizing the customers' right to set the performance and reliability they require. Next come the whole of the manufacturing company's design, production and test staff. Finally we have all the problems which beset the user who is striving for reliable operation. It is thus very difficult to find anyone who is involved in the project and not in some way concerned with its reliability. Clearly:

RELIABILITY IS EVERYBODY'S BUSINESS

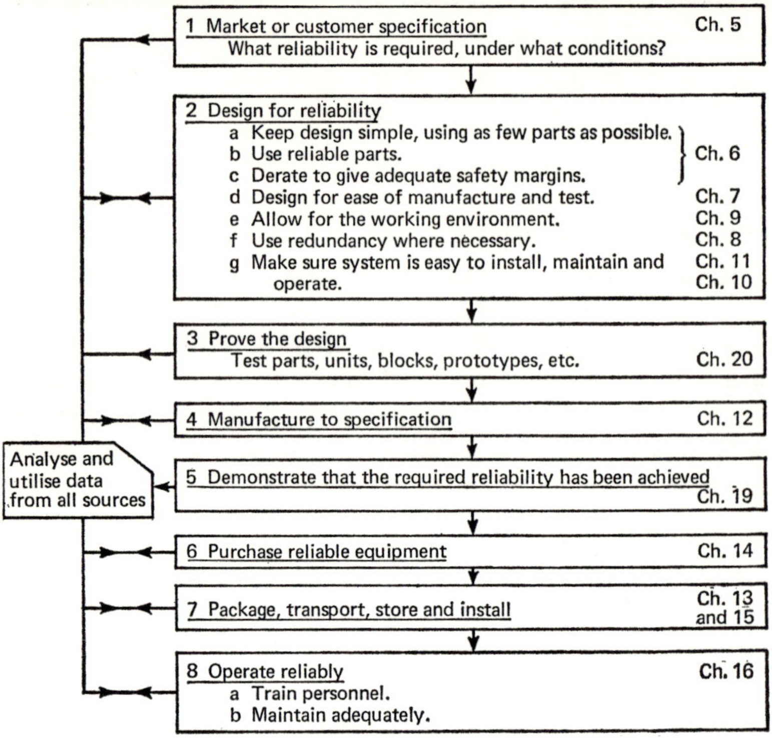

FIG 22.2(a). Reliability in a nutshell

22.3. coordination for reliability

Coordination for reliability therefore presents a major problem. Even the managing director of a company can only control the people it employs. He cannot control the customers. Similarly those who are responsible for the reliable operation of equipment seldom have any direct control over its manufacture, although they may be able to exert pressure on those who do.

This means that the role of the reliability manager must be one of coordination rather than direct control. He must ensure that all those who are responsible for the functions set out in Fig. 22.2(a) work together to achieve a satisfactory overall reliability. Thus the designer must be responsible for the reliability of his own designs, and the factory for making to specification. The reliability manager however is the one who encourages, and if necessary ensures, coordination between them.

The duties of the reliability manager may therefore include the following:

1 Coordination between departments, so that each is quite clear what its reliability responsibilities are.
2 Collection of reliability and associated information from all sources, analysing it for trends, etc. and then feeding it to the departments which need to make use of it.
3 Reliability demonstration tests.
4 Reliability measurements on prototypes, new materials and parts.
5 Reliability audits. These are sample checks, designed to ensure that everything is satisfactory from a reliability point of view.
6 Organization of reliability prediction as discussed in Chapter 21.
7 Training for reliability as discussed in section 22.4.

22.4. training for reliability

The reliability manager must ensure that his staff have adequate training and experience. Basic reliability theory and techniques can be studied at various technical colleges, but training in the reliability methods as used within an individual company are usually best done by an internally run course, possibly with outside help. In addition it will be necessary to see that other staff are reliability minded, and that people like designers, production engineers and industrial engineers are familiar with the basic reliability concepts which have a relevance to their work.

22.5. matching the solution to the needs of the problem

There is no single right way to solve reliability problems, and the reliability manager and his staff must study each carefully.

The following steps are nearly always necessary in some form or other:

1 Collect as much relevant information as possible.
2 Analyse it fully, taking care not to either miss something important or jump to conclusions.
3 Devise an appropriate solution. This might be exactly as it appears in a text book, but more likely it will be adapted to match the needs of a particular situation. We should have no hesitation about doing this, provided we are sure that our proposals are basically sound. The most satisfactory solutions will be characterized by:
a Their relative simplicity, in that they will contain no unnecessary frills or complications.
b The exact way in which they solve the problem in hand.
4 The proposed solution should be checked and tested to make sure it does all that is required of it.
5 Next it is implemented, and this will be all the easier if everyone can see how apt and appropriate it is.
6 Finally we shall use our experience on one problem to help us with the next, and so build up a bank of useful and practical reliability experience tailored to suit our own particular sphere of operations. It will be backed up by subsequent proof in service.

22.6. the relationship between quality and reliability

Quality is sometimes defined as 'fitness for purpose', and we can break it roughly into:

1 *Physical features*, e.g. whether an item has a satisfactory appearance, all its dimensions within limits etc.
2 *Performance*, i.e. whether it works correctly.

As we said in section 1.2, reliability is the probability that an item will perform as required, under stated conditions, for a stated period of time. Hence since performance is an aspect of quality, we might loosely say that reliability is the probability an item will retain its quality, under stated conditions, for a stated period of time.

Thus quality and reliability are very closely related, and it is reasonable to ask why as subjects they ever became separated. There are probably two main reasons.

1 Reliability is concerned with the probability of failure at some instant in time, and therefore its mathematics are quite different from that used by the quality engineer.
2 In practice quality control tends to impinge chiefly on the manufacturing unit, whereas reliability tends to centre on design. As we have seen, however, reliability and indeed quality as well have much wider implications than this, and should not be so restricted.

In suitable cases, therefore, there is a lot in favour of combining both subjects under a quality and reliability manager. This avoids pointless arguments which sometimes occur about whether the reliability department should incorporate the quality control department, or whether quality control should have the privilege of absorbing reliability.

22.7. specification AvP 92

AvP 92 entitled Specification of Quality Management Requirements was issued by the then Ministry of Technology in June 1968, and copies can be obtained from Her Majesty's Stationery Office. Its contents are mandatory on companies engaged on government contracts, but it contains a lot of sound advice which may be of use to other companies as well. Most of it is concerned with quality, but it has a few important points about reliability:

1 The contract should normally state the reliability requirements, and these should be derived from cost effectiveness, performance, reliability, time available, etc. to give an optimized programme.
2 A quality manual must be prepared and kept up to date by each company, and among a lot of information in it must be the number and grade of staff working on reliability.
3 Designers must be aware of their responsibilities for reliability.

This specification is quite comprehensive and should be studied by all reliability engineers.

22.8. reliability and other subjects

As reliability engineers, it is very tempting for us to think of reliability as a single, isolated and very important subject. Indeed with a bit of imagination, quality and reliability combined can be regarded as the two great subjects of which all other management techniques, such as work study, ergonomics, value analysis, etc. form merely a part. This line of thought is possible because man's knowledge is really a unity, which we divide into subjects for convenience of study. There are no clear boundaries between them, for each links and entwines with the others, and cannot be properly studied in isolation.

As reliability managers, we must be careful to take a balanced view. For each problem we need the best solution and so long as we achieve this it matters very little what we call the methods and techniques we use.

TABLE FOR SELECTING A SAMPLING PLAN FOR A RELIABILITY DEMONSTRATION TEST

Number of Failures Allowed	Reliability to be demonstrated: 0·70		0·75		0·80		0·85		0·90		0·93		0·95		Number of Failures Allowed
	n	$RP_A = 0·9$	*n*	$RP_A = 0·9$	*n*	$RP_A = 0·9$	*n*	$RP_A = 0·9$	*n*	$RP_A = 0·9$	*n*	$RP_A = 0·9$	*n*	$RP_A = 0·9$	
0	8	0·99	9	0·99	11	0·99	15	0·99	23	0·99	32	0·997	45	0·998	0
1	13	0·96	16	0·97	18	0·98	25	0·98	38	0·99	55	0·990	77	0·993	1
2	17	0·94	21	0·95	25	0·96	34	0·97	52	0·98	75	0·985	105	0·990	2
3	22	0·92	27	0·93	32	0·95	43	0·96	65	0·97	94	0·981	132	0·987	3
4	27	0·91	32	0·92	38	0·94	52	0·95	78	0·97	113	0·978	158	0·985	4
5	31	0·90	37	0·91	45	0·93	60	0·95	91	0·97	131	0·976	184	0·983	5
6	35	0·89	41	0·91	51	0·92	68	0·94	104	0·96	149	0·974	209	0·981	6
7	39	0·88	47	0·90	57	0·92	77	0·94	116	0·96	166	0·972	234	0·980	7
8	43	0·87	52	0·90	63	0·91	85	0·94	128	0·96	184	0·970	258	0·979	8
9	49	0·87	59	0·89	69	0·91	93	0·93	140	0·96	201	0·969	282	0·978	9
10	53	0·87	64	0·89	75	0·91	100	0·93	152	0·95	218	0·968	306	0·977	10

Number of Failures Allowed	Reliability to be demonstrated: 0·97		0·98		0·990		0·995		0·997		0·998		0·999		Number of Failures Allowed
	n	$RP_A = 0·9$	*n*	$RP_A = 0·9$	*n*	$RP_A = 0·9$	*n*	$RP_A = 0·9$	*n*	$RP_A = 0·9$	*n*	$RP_A = 0·9$	*n*	$RP_A = 0·9$	
0	80	0·999	120	0·999	230	0·999	460	0·9998	770	0·9999	1150	0·9999	2300	0·99995	0
1	130	0·996	200	0·997	390	0·999	780	0·9993	1300	0·9996	1950	0·9997	3890	0·9999	1
2	180	0·994	270	0·996	530	0·998	1070	0·9990	1780	0·9994	2660	0·9996	5320	0·9998	2
3	220	0·992	330	0·995	670	0·997	1340	0·9987	2230	0·9992	3340	0·9995	6680	0·9997	3
4	270	0·991	400	0·994	800	0·997	1600	0·9985	2670	0·9991	4000	0·9994	7990	0·9997	4
5	310	0·990	462	0·993	930	0·997	1860	0·9983	3100	0·9990	4640	0·9993	9280	0·9997	5
6	350	0·989	528	0·993	1050	0·996	2110	0·9982	3520	0·9989	5270	0·9993	10530	0·9996	6
7	390	0·988	590	0·992	1180	0·996	2360	0·9980	3930	0·9988	5890	0·9992	11770	0·9996	7
8	430	0·987	650	0·992	1300	0·996	2600	0·9979	4330	0·9988	6500	0.9992	14000	0·9996	8
9	470	0·987	710	0·991	1420	0·996	2840	0·9978	4740	0·9987	7100	0·9991	14200	0·9996	9
10	510	0·986	770	0·991	1540	0·995	3080	0·9977	5150	0·9986	7700	0·9991	15400	0·9995	10

1. This table applies to large batches.
2. $RP_A = 0·9$ means 'The reliability which would have an acceptance probability of 0·90'.

INDEX